OCEANIC BECOMING

OCEANIC BECOMING

The Pacific beneath the Pavements

....

Rob Wilson

DUKE UNIVERSITY PRESS DURHAM & LONDON 2025

Project Editor: Michael Trudeau
Designed by Matthew Tauch
Typeset in Warnock Pro by Westchester Publishing Services

Library of Congress Cataloging-in-Publication Data
Names: Wilson, Rob, [date] author.
Title: Oceanic becoming : the Pacific beneath the pavements / Rob
Wilson.
Description: Durham : Duke University Press, 2025. | Includes
bibliographical references and index.
Identifiers: LCCN 2024028356 (print)
LCCN 2024028357 (ebook)
ISBN 9781478031475 (paperback)
ISBN 9781478028260 (hardcover)
ISBN 9781478060468 (ebook)
Subjects: LCSH: Ocean and civilization. | Climatic changes—Social
aspects—Pacific Area. | Globalization—Environmental aspects—
Pacific Area. | Marine resources conservation—Economic aspects—
Pacific Area. | Sea in literature.
Classification: LCC CB465 .W55 2025 (print) | LCC CB465 (ebook) |
DDC 320.1/2091823—DC23/ENG/20241216
LC record available at https://lccn.loc.gov/2024028356
LC ebook record available at https://lccn.loc.gov/2024028357

Cover art: Illustration by Matthew Tauch, from a photograph
by Rob Wilson.

CONTENTS

Acknowledgments   vii

INTRODUCTION · Pacific beneath the Pavements · *Toward a Blue Ecopoetics of Oceanic Belonging*   1

I   WORLDING PACIFIC POESIS

ONE · Becoming Oceania · *Ecopoetics across the Planetary Pacific Rim, or "Walking on Water Wasn't Built in a Day"*   31

TWO · Worlding Asia Pacific into Oceania · *Concepts, Tactics, and Transfigurations inside the Anthropocene*   51

II   WORLDING THE PACIFIC RIM

THREE · Toward a Blue Ecopoetics · *Worlding the Asia Pacific Region into Figurations of Oceania at Monterey Bay*   71

FOUR · Migrant Blockages, Global Flows · *Worlding San Francisco in a Global-Local and Transoceanic Frame*   92

III   TRANSPACIFIC CONJUGATIONS: UNMAKING AND
REMAKING WORLDS

FIVE · Under a Golden Gate "Mushroom Cloud" · *Urban Space, Ecological Consciousness, and the Pedagogy of Blue Conversion*   111

SIX · Hiroshima Sublime · *Trauma, Japan, and the US Asia Pacific Imaginary*   126

SEVEN · Waking to Global Capitalism and Oceanic Decentering · *Reworlding US Poetics across Native Hawai'i and the Pacific Rim*   141

EPILOGUE · Transplanted Poesis · *Writing Oceania and the World*   161

Notes   167

Bibliography   197

Index   219

ACKNOWLEDGMENTS

The premodern Japanese poet Matsuo Bashō is rumored to have advised his fellow haiku questers, "Do not seek to follow in the footsteps of the men [or women] of old; seek what they sought." In the course of writing this transoceanic book laden with lifelong scholarly and poetic commitments, sites, and places from coastal California to Hawai'i, Oceania, Taiwan, South Korea, Hong Kong, and Japan, some crucial mentors have departed. Among them, three haunting and "aftering" figures of Asia Pacific renown abide—namely, Masao Miyoshi, Arif Dirlik, and Kenzaburō Ōe. I continue, in my own belated way, to "seek what they sought" as scholarly and writing ethos and as world-making obligation.

Over these years and shifting contexts I have continued to be influenced by, and indebted to, two long-lasting journals, *boundary 2: An International Journal of Literature and Culture* and *Inter-Asia Cultural Studies*, and the distinguished, prodding, and supportive editors at the head of these cultural and political collectives, Paul Bové and Kuan-hsing Chen. Affiliated scholars such as Donald Pease, Lindsay Waters, Jonathan Arac, Reynolds Smith, Beng-huat Chua, Meaghan Morris, Colleen Lye, Soyoung Kim, Ping-hui Liao, Kim-Uchang, Hsinya Huang, Andy Chi-Ming Wang, and others have been a cocreative part of this crucial configuration. These figures have supported special issues over these years and geohistorical contexts and provoked shifting discursive turns of rewording vision for this author, and I remain grateful.

Other scholars working more centrally in the critical oceanic frame, particularly Steve Mentz and Elizabeth DeLoughrey, as well as the late Paul Lyons, have also proved lasting influences and enablers of shifting tropes and turns. Writers and artists as diverse as Gary Pak, Steve Bradbury, Susan Schultz, Joseph Puna Balaz, Juliana Spahr, Maxine Hong Kingston, Michael Ondaatje, and Jake Thomas also figure in this Asia Pacific world-making, and oceanic trajectory enacted herein.

The University of California at Santa Cruz has been my institutional home since I moved from the University of Hawai'i at Mānoa in 2001. Scholars and writers here, including Christopher Connery, Karen Tei Yamashita, Susan Gilman, Kirsten Silva Gruesz, Ronaldo Wilson, Micah Perks, and Christine Hong as well as transdisciplinary figures such as James Clifford and Donna Haraway, have influenced these transformations of an Americanist-trained scholar, creative writer, and teacher into a more global-local, ecological, and Asia Pacific one, and I remain grateful. I also need to invoke the support and influence from the Center for European and American Studies at Academia Sinica in Taipei, Taiwan, and its affiliated universities and scholars there. Serena Chou's environmental vision as activated around the Farm for Change project has enabled ecopoetic works across "Asia Pacific becoming Oceania," and the support of the institute's leader, Norman Teng, as well as of the National Science Council, has proved helpful across these transpacific projects, years, sites, and distances.

Duke University Press has been the crucial outlet for nearly all of the books and coedited collections I have published from the 1990s to the present. For the shape and totality of this capstone work, *Oceanic Becoming: The Pacific beneath the Pavements*, I am grateful for the input, advice, and support from Ken Wissoker, Michael Trudeau, and Kate Mullen, as well as from others behind the discursive scene, such as the design team and copyeditor. As Donna Haraway has long urged, such modes of creating and world making are always a "creating with" (sympoetic) undertaking of interconnection and entangled work and care.

Pacific beneath the Pavements

Toward a Blue Ecopoetics of Oceanic Belonging

Without an understanding
of the world system and the sea as the space of commerce
it is hard to integrate that other
most important fact of our era. Pirates [riots].
· JOSHUA CLOVER, *Red Epic*

Occluded beneath the oblivious pavements of prosperous global cities such as San Francisco, Seoul, Taipei, and Hong Kong—not to mention Euro-distant Berlin and Hamburg (with their ex-colonial ties to Western Samoa and North Sea access to world oceans) or Paris (hub of the French Pacific in Tahiti, site of toxic nuclear testing)—the ebb and flow of the Pacific Ocean abides as a hydraulic system, source of life and breath, material resource, planetary nexus, and site of plasticene and industrial oil spill and waste: Anthropocene reminder.[1] This mighty Pacific Ocean, as Herman Melville prefigured by the mid-nineteenth century of American adventurism, belts, links, and zones the whole world into geospatial integration— what he called, one hundred years before the environmental science of Rachel Carson, "the terraqueous globe."[2] Gazing into the future, Lawrence Ferlinghetti, San Francisco poet and founder of City Lights Press, could see this disturbed ocean overtaking city embankments, streets, and side-walks that have made North Beach a dwelling place for art, commerce, culture, coffee house, and community. "[San Francisco's] all going to be underwater in 100 years or maybe even 50," Ferlinghetti warned in 2017. "The Embarcadero is one of the greatest esplanades in the world. On the weekends, thousands of people strut up and down like it's the Ramblas in Barcelona. But it'll all be underwater."[3] Venice might be worst-case scenario of a world city threatened by oceans. As the Asiatic rises above and below city canals, streets, and foundations, the technologies of an adjustable oceanfront wall and pumped-in underground saltwater can-not forestall threats of urban disaster.[4] The ocean is not just under this city but inside it, over it, rising above its walls and paved streets, as it is even in a smaller coastal city such as Santa Cruz, California, where I am writing this.

Ferlinghetti knew as a World War II naval submarine officer defend-ing against German U-boat attacks at Normandy and Norway, and as an aftermath witness to the Nagasaki and Hiroshima bombings in devas-tated Japan, that the Pacific Ocean (like the Atlantic) was never that *pacific* as an ocean site in the modern world system. This proved so across world history, despite Ferdinand Magellan's ill-fated trope baptizing "el Mar Pacifico" while traversing myriad archipelagos in 1521 for the Portuguese and Spanish Catholic Church. Ferlinghetti refused to defend the use of nuclear weapons in Japan to induce capitulation in August 1945. "In that instant," Ferlinghetti admitted years later, "I became a total pacifist." These American wars across the Pacific, as later in Korea and Vietnam, turned Ferlinghetti from US militarism and toward pacifist activism, a poetics of resistance, and the community he would help build around City Lights Bookstore in the transpacific city of San Francisco after the war. "It was a monstrous, racist act, the worst the U.S. ever committed," Ferlinghetti contended, looking back on his career. "Had the Japanese been white-skinned, those bombs would not have dropped."[5]

Unstable atmospheric currents of El Niño and La Niña have made climate patterns across the Pacific Rim more threatening with flood and drought, disruptive events on a seasonal if not daily basis. This Pacific Ocean (like the Empire-laden Mediterranean, Atlantic, Adriatic, and Indian oceans) has become filled with history, struggle, bloodshed, exploitation, ideological division, and projection, from Vietnam and Manila to Guam, Fiji, the Solomon Islands, Jakarta, and Pearl Harbor. As origin and frontier, this Pacific becomes diversely figured and transfigured into primordial mother water and cargo lane, hydraulic matter and trans-species element, as "blue" consciousness and as integrating biosphere, alpha, and omega off the coasts of the Americas and Asia, as if some dream of trans-indigenous Oceania, or infrastructural apparatus and commercial ideologeme.[6] At deeper levels of urban-oceanic interconnection, this blue Pacific has darkened, blackened, and reddened with plastic waste, dying coral reefs, proliferating jellyfish, toxic agribusiness remainders, radioactive traces, raw heat, record-breaking typhoons, extreme weather events.[7] As in the North Atlantic city of Hamburg, with its recurrent flooding and "waterlogged history" moving from the medieval struggles of the Hanseatic League integration down to the financial terrains of its becoming a cosmopolitan global hub for the European Union, as Kay MacFarlane reminds us, "both the city's prosperity and its precarity have been closely tied to its marshland geography."[8] Germany, despite its continental impact across the twentieth century—if not earlier in the imperial scramble for Africa and expansion eastward and westward—remains a deeply oceanic nation.

In the wake of hegemonic globalization and techno-integration, this world-ocean Pacific has become all but covered and troped over with pavement tracks and containerized shipping routes, garbage heaps, real estate schemes, DDT waste, "South Pacific" romance, storm, tsunami, Google warehouses, radioactive specters, Godzilla, and worse. Still, as I argue here, this "Pacific becoming Oceania" reflects primordial longings for bio-marine sustenance and environmental endurance within and across this ocean. In effect, this ocean commons of the Pacific needs to be figured as both *peril* and *promise*, articulated at social and ecological levels that trace damage and potentialities, as elaborated in later chapters reflecting transpacific contexts and urban sites of dwelling, connecting, resisting, and world belonging, from coastal California across the Pacific to Oceania, Asia, and the Pacific Rim.[9] Since subscribing to the exclusive economic zone's maritime doctrine of two hundred nautical miles of offshore boundaries in 1982—as Brian Russell Roberts argues, while decentering continental-centered frameworks of the United States into an "archipelagic" nation of oceanic-island fluidity and "border water" remaking—"The United States Is an Ocean Nation."[10] Cities along the Pacific Rim and at the Oceania or Asian edge of oceanic becoming—from Auckland, New Zealand, to Hong Kong and Kaohsiung, Taiwan—embody the precarity (peril) and prosperity (promise) of this global-local positioning across this world-ocean Pacific. When we gaze, as Ferlinghetti did from the Embarcadero walkway along San Francisco Bay, across to Hiroshima and the Solomon Islands, do we project the precarious history as when he transfigured coastal California at an Anthropocene end?

How do we figure forth hidden yet ever present links to this mal-troped, occluded, circulatory, endangered, un-pacific, and archipelagic Pacific; this blue *pharmakon* (remedy and toxin) of ocean water that embodies both healing and poisoning presence? To far-flung coastal citizens whose prosperity generates effects gone awry, such as the Great Pacific Garbage Heap and the oil spills of Santa Barbara, as well as the nuclearized overflow of Fukushima, this is what the Maori scholar Alice Te Ponga Somerville unpacks as "the (American) Pacific you cannot see."[11] What tactics or figurations of world-making *poesis* are needed to bring into critical/poetic articulation urban, coastal, island, and archipelagic links to this "Great Ocean" of epochal transformation that took far-flung historical effect across the globe? How do we grasp layers of Spanish, Portuguese, Dutch, British, French, American, Chinese, Japanese, and German Enlightenment contact and colonial entanglement, not to mention the scientific integration of the Pacific into the world capitalist system of Greenwich Mean Time, as well as the dynamism of global-capitalist modernity that now rules as naturalized norm, if not ethos and aesthetic?[12] To invoke a Japanese-influenced haiku by African American novelist Richard Wright on this oceanic presence as a breathing world body, filling each instant with what the English post-priest Gerard Manley Hopkins called (in a Roman Catholic register) the "world-mothering air" of blue sustenance and planetary nourishment:

> The ocean in June:
> Inhaling and exhaling,
> But never speaking.[13]

Still, as Elspeth Probyn urges of this waste-transforming and nourishing element of the world ocean that is around, above, and inside us, "Eating the ocean: We do it every day, often without knowing it."[14] Food resources link sites, shores, and cities through fishing, farming, consuming, polluting, wasting, bioprospecting, and consuming its oceanic creatures. From Sydney, Honolulu, Auckland, Seattle, Vancouver, San Francisco, Los Angeles, Tokyo, and Seoul along the Pacific Rim to Glasgow, Dublin, Oslo, and London across the Northern Atlantic, saltwater ties and oceanic affects become ingested, processed, embodied, and all but forgotten in urban settings upscale.[15] *Worlding poesis*, as an actively critical and constructive process, might help to figure forth such responsible modes of world be-longing, projecting tactics of dwelling together and multiscalar commun-ing. These can happen in the "blue humanities" of oceanic and archipelagic recognition via revealing land-sea interconnections that spread across material-semiotic binaries that come to be called (by Donna Haraway, Anna Tsing, Edgar Garcia, and others) modes of trans-species *sympoesis*— that is, poesis as a *making with* coral reefs and marine creatures, snails and whales, sharks and quarks, macro and micro forces, signs on Earth and in the sky that indigenous peoples long cultivated and lived beside.

These ties, signs, linkages, bonds, and balances are threatened by late capitalist patterns of consumption, extraction, extinction, and all-too-poisoned Apple production chains of labor and surplus value from Silicon Valley back and forth across the Pacific to Greater China, Europe, and the world.[16] Linked to and living with this *oceanic commons*, at least here beside and across the Pacific Ocean in coastal California, has all but mod-ulated from the "Asia/Pacific" hub of geo-economic dynamism and the preferred US-hegemonic trope of the "Pacific Rim," which came to promi-nence in research sites and policy projects in California across the 1970s, 1980s, and 1990s as Chris Connery has mapped as global-capitalist neces-sity. This region has gone on changing, if not booming and busting, into the commercial, air traffic, and cultural transit of some open-ended, unregu-lated "transpacific" flow—one that Global China nonetheless threatens (as the ill-fated Trump regime abolished "transpacific" trade contracts and environmental commitments that the Obama administration had ratified from Paris to Kyoto) to integrate into some post-Bandung neo–Silk Road of the "Great Chinese Dream." With or without post-COVID Trumpian affiliation, racial stigmatization, and tariff blockage, Global China (centered on state one-party hegemony in the People's Republic of China [PRC]) projects, implements, and funds this *transpacific* vision as a One Belt, One Road unity spread across world oceans, as well as across distant continents, into one infrastructural investment and com-mercial—if not geopolitical or cultural—dominion.[17] Chinese expansion is often critiqued as a top-down threat to the archipelagic United States in and across the Pacific, even as this rebordering US nation continues to shift from seeing itself as what Roberts terms a "majority-continent to a majority-ocean" people moving across postwar hegemony.[18]

As a global nation seeking to install an alternative hegemony across the Pacific via the southward oceanic telos it has long projected histori-cally, as did Japan in the buildup to World War II around Pearl Harbor and Guadalcanal, this ocean-challenging China of naval, financial, and technological might has already built four hundred buildings on just one of its man-made islands constructed on the Subi Reef in the South China Sea, amid the contested Spratly Islands off the coasts of the Philippines, Malaysia, and Vietnam.[19] As if to signal the international instability of this Pacific region and the waning naval hegemony of the United States as measured against the rising presence of this maritime China, the US Pacific Command (PACCOM), with headquarters at Pearl Harbor, has been renamed the US Indo-Pacific Command Center, as if India has become the key affiliated naval player in the region. The Trump administration turned what was called for three or more decades the "Asia Pacific" region into the "Indo-Pacific" to signal that India, not China, bookends the US military hegemony across this oceanic region—"from Hollywood to Bollywood, from polar bears to penguins," as US Secretary of Defense James Mattis phrased this shift of names, tropes, and visions from the "American Pacific" in Hawai'i.[20] As the *Economic Times* of India warned, this renaming reflects "a largely symbolic move to signal India's importance to the US military amid heightened tensions with China over the militarisation of the South China Sea."[21] India does have a deeply oceanic history via the seas of Arabia and India, as well as across the Pacific, but becoming an international military or naval power of global consequence is not yet one of its primary global aims as it at least rises to BRIC (Brazil, Russia, India, China) economic recognition.

This Pacific Ocean—laden with worldly history, memory, community, conflict, and trope—not only becomes a "physical space" of oceanic materiality and physical exchange; it is once again projected as a "horizon of possibility" linking the cultures between Asia and across the United States and its Pacific claims and flows. Hua Hsu describes this "floating" transpacific of dream and scheme as some enduring Chinese nexus of "transience, motion, and flux" from the Gold Mountain days of building railroads to the containerized shipping power of the Pacific-facing China of Nanjing, Shanghai, Guangzhou.[22] An action movie such as Rupert Sanders's *Ghost in the Shell* (2017), based on Mamoro Oshii's Japanese anime classic from 1995, captures this transpacific flux in its post–*Blade Runner* ambient mix of Hong Kong, Shanghai, Tokyo, Long Beach, and Los Angeles into some post-Orientalist fusion of dream, machine, lust, flesh, and scheme. No less inventively as transpacific cinema, Wes Anderson's translingual (Japanese and English) film *Isle of Dogs* (2018) gets at this brave new world of machinic flesh and trans-species critters (it's a dog-eat-dog world, as we say of capitalist mores from Tokyo to San Francisco and Wall Street) toxically, in a post-Hiroshima and post–Japanese internment modality, to reveal the cruelty, injustice, and nuclear horrors inside this island-hopping dream spread along the Pacific Rim.[23]

Jasper Bernes has portrayed this dystopian captivity of ocean beaches under the pavements of late capitalist Los Angeles by excavating (in post-Situationist terms) sub-pavement layers of concrete urban enclosure in the devious haiku "documents" poem from *Starsdown*:

> Under the parking lot, the beach.
> Under the beach, the parking lot.[24]

Beneath the concrete, via urban synecdoche, there is still more concrete stifling ocean flow and occluding Pacific watersheds along the once thriving Los Angeles River. Beneath world ocean floors, more hopefully, Australian marine scientists are coming to discover there may exist resources of freshwater aquifers that could help Earth dwellers to deal with heat droughts and rising currents that are being projected for 2030 by the United Nations.[25]

Our anxiety over late capitalist urban enclosure, political stasis, and anti-utopic demoralization may sound a bit like this rueful refrain from a Parisian tweet expressing how post-1968 revolutionary energies have passed us by as possibility: "Sous les pavés: la plage. Sous la plage: plus des pavés. Sous ces pavés: une rue. Pavé par une plage."[26] Still there is another story to tell—what Dale Pendell, in his biopoetics as trans-species bonding with flora and fauna and alchemical transformation of earth and planet, calls the "understory" that needs to be given a second chance before end-times are here.[27]

Pacific Awakening in the Berlin Anthropocene

While living and working in various cities inside and along the Pacific Rim since the mid-1970s I came to cosmic realization of our world entanglement, oceanic occlusion, and urban obliviousness. This oceanic epiphany took place at the Haus der Kulteren der Welt (House of World Cultures) in the postmodern city of Berlin in 2013, where I was speaking as part of an international conference (with the Pacific writers Albert Wendt, Juliana Spahr, Reina Whaitiri, and Axel Hein, among others) on what the museum planners called "The Invisible Pacific." The program used thematic slogans of historical misinformation, such as "Circling the Void: Long Night of the Pacific," to capture German audiences' attention. The conference was also part of a multicultural "Wassermusic" (Water Music) program filled with diverse works of Asian Pacific multiplicity at the same time that this progressive institution hosted a vanguard series of scholarly talks and "museum without walls" interventions into canals, plazas, and walkways of Berlin to awaken public awareness to what was coming to be called here, as elsewhere, the Anthropocene.[28]

Given this large-scale enframing into a global *telos* of world extraction and extinction, island ecologies and spaces are all the more threatened by world-capitalist and carbon-driven industry since the Industrial Era and accelerated by the atomic explosions at Hiroshima and Nagasaki, as well as at the Bikini Atoll across the Pacific Ocean.[29] Berlin, this urban, hip, and cosmopolitical post-1989 city, felt close to this endangered Pacific of toxic coral in the Great Barrier Reef and the melting icebergs and heating waters of the Arctic region having an impact on cities and farmland from India to Peru. At least we were becoming aware of distant islands such as Tuvalu disappearing back into "some void Pacific" in which the Pacific islanders will live on only as a website in a far-flung diaspora from their island home.[30] It no longer makes sense, as during earlier phases of urban globalization (such as the buildup for the 1939 San Francisco World's Fair), to figure San Francisco or Los Angeles as coastal cities of frontier-space California staring into "the void Pacific," or just second-rate urban Asia models as distant belated copies of urban splendor in the empire cities of Paris, Vienna, and Rome.[31]

Sous les pavés, la plage! (Under the paving stones, the beach!): Can this geopolitical call to psycho-geographical disruption reawaken subterranean energies latent inside or beneath the city form? This revenant slogan from the streets of Paris in 1968, as well as from Berlin in 1989, will stand for what I call here (more broadly framed as reworlding synecdoche) "the Pacific beneath the pavements." This slogan, in effect, calls out for urban citizens to disrupt spectacles of neoliberalism and everyday urban space; to challenge commercial normalcy at the arcade and mall; to intervene in the political stalemate and mass-mediated discourse.

This call for geopolitical eruption scrawled on walls of demonstration-ridden Paris endures in cultural studies as a post-Situationist slogan of resistance to spectacles of capital domineering over urban streets as commercial traffic and consumption. McKenzie Wark uses this slogan

8

for *The Beach beneath the Street: The Everyday Life and Glorious Times of the Situationist International*, wherein he invokes Situationist tactics (*derive, detournement*, to use situation as intervention) for urban transformation and creative use.[32] If resituated as a tactic of worlding politics and oceanic poetics, that same "submarine" street slogan from Paris '68 can provoke post-Situationist critiques, surrealist dreams, and interventions into urban life as domesticated under capital. Wark elaborates: "One [dynamic of urban intervention] was communist, and demanded equality. The other was bohemian, and demanded difference."[33]

This reframing of urban prosperity and complacency inside the Anthropocene makes urban-oceanic reckoning and worlding tactics along the Pacific Rim and across Oceania all the more urgent as Earth dwellers become urban citizens on the endangered planet.[34] As Nick Mirzoeff urges, the "politics of seeing" inside the global city of capitalist expansion has become foreclosed, suggesting a failure of social imagination and collective futurity: "There were once eight million stories inside the *Naked City* (1948), but now there's only one: the endless rise of the one per cent for the one per cent."[35] Class polarization, securitized segregation, resource extraction in distant sites across orders, habitat destruction, and community uprooting via gentrification, as well as various forms and modes of environmental plunder (sublimated into plastic or greenwash disguise), have become givens of this "global city" as the dominant bioform from Beijing and Hong Kong to San Francisco, Berlin, Honolulu, and Rome. Such a habitus seems all but immune to political contestation or any surrealistic dream projecting alter-reality.

Martin Jankowski, the German novelist host for our panels at the House of World Culture, reminded participants, as we talked beforehand about the "Invisible Pacific" event to be conducted as a chain-linked interview of writers, scholars, and marine biologists from the Atlantic to the Pacific, that the earlier German dream of "Visafrei bis nach Hawai'i" (visa-free travel all the way to Hawai'i) had served as a mobilizing slogan for mass demonstrations in Leipzig (East Germany) and West Berlin. These urban demonstrations resonated at the deepest level with the hunger for global travel and post-Beat longing that (among other social forces) helped lead to the fall of the Berlin Wall and to the opening of Germany to the dynamism of globalization that the rebuilt Berlin and open Brandenburg Gates now stand for inside the European Union. The beaches and saltwater of this remote Pacific Ocean, then as well as now, can call out to be reckoned with by a self-reinventing Europe (as across the transnationalizing

Americas) beyond the reign of urban capitalist modernity. These cities have seen movements and generations contesting the geo-ecological impact of rising tides and shifting coasts, from the North Atlantic of beachtown New Jersey to the monster surfs of coastal Northern California Pacific and tidal-threatened Pacifica, not to mention the coastal real estate of Florida and North Carolina that only the most climate-denying anti-science forces inside can continue to ignore as extreme weather.

Visa-free travel had become the anti-Wall German dream of visa card travel to white beaches and eternal summer, such as those in the islands of Hawai'i, Tahiti, and Guam, far from urban Berlin and Dresden. This is how a blog entry recounts those East German contexts to which I have been alluding, which led to the mobilizing call and that transpacific slogan "Visafrei bis nach Hawai'i":

> At about 7.30pm on 9 November 1989 a spokesman for the GDR [German Democratic Republic] government said on television that all restrictions on foreign travel by GDR citizens were lifted with immediate effect. Vast crowds gathered near the Brandenburg Gate (on the East Berlin side) and demanded their right to travel to West Berlin. After initial bewilderment, the East German border guards let them cross. (This had been preceded by several weeks of demonstrations demanding, among other things, freedom of travel. "Visafrei bis nach Hawaii"—"visa free travel, all the way to Hawaii" had been one of the key slogans.)[36]

The pressure of such mass demonstrations in Berlin and Leipzig (which the younger Berkeley- and Paris-influenced Martin Jankowski had participated in as a protesting student inside the GDR) led to the fall of the wall on November 9, 1989, marking the fall of the socialist GDR regime and the opening of Germany to forces of mobility, democratization, liberalization, and what would become a more transnational Europe of open borders and cosmopolitical contestation.[37]

As Tom Brislin recounted in pages of the *Honolulu Advertiser* on the tenth anniversary of the Berlin Wall's dismantlement in 1999, "German journalist Kristin Schonfelder recalls shouting the slogan as a nineteen-year-old university student with thousands of other candle-holding marchers, pushing back the darkness of the streets of Leipzig in the weeks before the crumbling of the Berlin Wall. 'It means "Without a visa to Hawaii,"' Schonfelder says. 'We wanted freedom to travel as far as we wanted, and Hawaii was the farthest away of any place we could imagine. And now,

10 years later, I am here. I never would have thought it possible.'"[38] Resonant with the Native Hawaiian struggle for political sovereignty and the right to sacred mountain access and oceanic custody of its island environment linking ocean and land, this Hawai'i of Oceania is not all that far from the geopolitical energies and environmental dreams of Berlin, Tokyo, Washington, Los Angeles, or the transpacific city of San Francisco—that Bay Area city of "blue mind" orientation "surrounded by water on three sides."[39]

To contribute to what is coming to be called the "blue humanities" of oceanic and archipelagic studies, we still need to extend world making (as the German expression has it) "ins Blaue hinein": into those blue depths, entanglements, and planetary, as well as localized, mixtures of the blue (healing) and red (endangered) ocean.[40] The ocean is felt, lived, and worlded into—*pharmakon*, that figure crucial not only to Plato and Jacques Derrida but also to the emergent biopoetics I am invoking here around the comparative poetics of metamorphosis, conversion, transfiguration, environmental imagination, and trans-species belonging that Norman O. Brown and Dale Pendell called the becoming multiple vocation of worlding into sanctified presence as "love's body."[41]

Break on through World Divisions

If we could get beyond urban confinement and coastal capture to so-called Oceania here being figured forth as ecological future and planetary commons, if the cultural-political front could only more collectively *break on through to the other side* of the Anthropocene, as the Doors conjured in the heady protest and Blakean days of 1967, then these urban pavements might be broken up, dismantled, collaged, even used as weapons cast against the forces of state repression and global-local containment that Allen Ginsberg memorably evoked in *Howl* as the San Francisco and Market Street figure of Moloch. Such modes of resistance I had lived through as a generational tactic, not only in the Berkeley demonstrations of the late 1960s, where I was an undergraduate in English at the University of California (UC), Berkeley, and a fringe member of the Haight-Ashbury in San Francisco, but also in the worker- and student-led mass street demonstrations daily marching against the South Korean military government of President Chun Do-won when I was a Fulbright Professor of American literature at Korea University in Seoul from 1982 to 1984.[42] Such energies of resistance have linked the Occupy movements in the Bay Area port city of Oakland and other cities across Northern California into a formation that Marxist critics from UC Davis, UC Berkeley, and UC Santa Cruz have called a Red Triangle struggling against the forces of privatization and debt bondage.[43]

This eruptive "beach" of utopic waters and sands living under locked-in pavements and ports of modern Paris was well known to post-'60s Californians, for whom a "subterranean homesick" politics called out for antiwar unrest, urban refusal, social movement, and human rights demonstrations as merged with post-Beat tactics from the libidinal bohemian culture of the Beats, the Doors, Sun Ra, and the Hendrix trio and the caustic visionary allegories of Bob Dylan. Invoking "the beach beneath the pavements"—in Paris then; Berlin later; or Seoul, Oakland, New York, and San Francisco—can stand now for reawakening what has been called "a subaltern vitality, the control of something unruly, the dominance of nature, and a possible return of the repressed . . . , [as well as] a new kind of social imagination, a right to view the city as a space of democratic possibilities, a social geography of freedom," as Benjamin Shepard and Greg Smithsimon contend in portraying New York City's urban domination of pleasure, community, and resource in *The Beach Beneath the Streets: Contesting New York City's Public Spaces*.[44]

Walter Benjamin once theorized those "ragpicker" forms of consumer capital to be excavated through Baudelaire's urban sidewalks and iron-and-glass Arcade settings in Paris as so many "dialectical images" in fleeting illuminations of the sacred within urban settings in *Les Fleurs du Mal*. "The Paris of [Baudelaire's] poems is a submerged city," Benjamin contends, "more submarine than subterranean."[45] Indeed, any modern city form itself depends on abundant water supplies for its conurbation buildup, from reservoirs to ports, from lakes and rivers to oceans, for livelihood and survival, as I discuss when examining the transpacific and Silicon Valley "imperial city" of San Francisco not just as a "subterranean" Beat site but also as a "submarine" cultural-political space of far-flung watersheds; rivers from the Sierras and Nevada; and abundant Pacific nourishment at Fisherman's Wharf and in the gourmet-ghetto bistros of Berkeley, Marin County, and Napa Valley. Sophie Gonick has urged seeing global cities such as San Francisco as a "city of contradictions" wherein progressive politics and the nineteenth-largest economy in the world confront high-tech hegemony, housing scarcity, class polarization, and the threat of rising sea levels: "In a moment in which the populist right wing is ascendant globally, cities can serve as beacons of hope with robust local articulations of democracy, alternative modes of politics and inhabitation, and popular imaginaries of the good life against the revanchism of many central governments."[46]

The hope remains that a subterranean/submarine "psycho-geography" of dis-alienated being will one day be released from beneath the pavements and ports of the global city into everyday domains of pleasure, freedom, abundance, release, festival, disalienation, dream, and what Brown called the transfiguring release of "metapolitical" energies, tropes, and libidinal movements from below.[47] Such a "metapolitical" Pacific Ocean summoned from beneath urban pavements and this "oceanic consciousness" of unity would stand for more than a world multiculturalism that, wrongly configured, would reaffirm the superiority of Europe in literature and art amid the ongoing "de-Europeanization" and decentering of global culture, as Masao Miyoshi notes in his portrayal of the Documenta X art show at Kassel, Germany, in 1998.[48] Facing world environmental crisis, Miyoshi invoked what he termed a necessary *planetary* turn "to nurture our common bonds to the planet."[49]

Such a planetary-framed Pacific, situated in the emergent world of the "blue" environmental humanities, would stand for more than just another "'Polynesianism' without Polynesia" that Jean-Didier Urbain sees in European fantasies of beach and oceangoing from the era of Daniel Defoe and Jules Michelet through the 1950s. This is what Urbain calls a *"robinsonnade"* escape to the Pacific as site of world forgetfulness. In this long-wrought European "aesthetics of the void, that underlies the vacation conquest of the seacoast," he argues, beachgoers would pacify the savage ocean and strip the middle-class tourist sites of native presence or lingering threats of phobia, terror, and slime.[50]

Pacific Transfigurations

Awakening this paved-over Pacific and world oceans beneath terrestrial urban streets of capitalist development must evoke into "oceanic consciousness" not just dreams of metapolitical unification, promise, and release, but also more catastrophic implications, peril, and threat: the ocean as space of global warming, methane gas, decimated coral reefs, nuclearized tsunami waters, and disappearing islands and glaciers, not to mention the mount-ing Great Pacific Garbage Patch of submarine waste lurking from coastal California to the waters of Hawaiʻi, Japan, and coastal China. The Pacific beneath the pavements of Europe and the Americas embodies not just the eroticism of the bikini (a postwar bathing brand of erotic titillation tellingly registered in 1946), but also the catastrophic impact of US nuclear testing across the Bikini Atoll, wherein sixty-seven nuclear weapons tests took place between 1946 and 1958. As the Native Pacific scholar and poet Teresia Kieuea Teaiwa has delineated, this "military-tourist" conjunction brings the world of Parisian eros (*the urban seen*) together with that of Pacific islander trauma, nuclear displacement of islanders, and radioactive slow death (*the urban unseen*) on the beach.[51]

At a level of urban consumption, Hong Kong's ties to world oceans occur not just through world-port shipping but also through abundant seafood consumption: having depleted local marine life and depen-dent on imported seafood, its citizens consume four times as much as the global average per capita. This taste for ubiquitous seafood eateries draws on the overfishing of unsustainable oceanic resources, "driving the collapse of the world's ocean fish stocks and edging many types of fish toward extinction," especially across the Pacific.[52] Using William Bur-roughs's tactic of desublimating cutup, we might ask (as Burroughs did of Ginsberg), "If we cut you up, Hong Kong, who would we find inside?" The answer that lurks beneath the streets and ports of Hong Kong (or the wealth of London, Tokyo, Shanghai, or Berlin) is, we can and will find the nexus of *world oceans* sustaining the world city with world-mothering air and planet-circulating waters.

From the time of Shakespeare, if not earlier in biblical hermeneutics, as the "blue cultural poetics" scholar Steve Mentz has argued, the ocean has figured in the Western imagination as both "a challenge to empirical understanding on the one hand, and seeing it as a divine Absolute, a God space that humankind can see but not understand on the other."[53] The allure of the transhuman ocean as space of God mystery, metamorpho-sis, death, and self-transformation—haunting magical-realist novels such as Yann Martel's *Life of Pi* (2001), David Mitchell's *Cloud Atlas* (2004), and Chang-rae Lee's dystopic *On Such a Full Sea* (2014), not to mention oceanic-based poems of indigenous world remaking such as the Maori poet Robert Sullivan's *Star Waka* (1999) and the Native Hawaiian poet Brandy Nalani McDougall's *The Salt Wind / Ka Makani Paʻakai* (2008), and ecopoetic works such as Juliana Spahr's *Well Then There Now* (2011), discussed in later chapters—has figured as geomaterial substratum and medium of world-altering modernity.

Oceans are connected not just to the movement of capital, peoples, and goods, but also to the material formation of city watersheds, international polities, war, demonstration, migration, weather, utopic longing, dream. From Melville's catastrophic novel *Moby-Dick* to Jules Verne's enchanted *Vingt Milles Lieues sous les Mers* (*Twenty Thousand Leagues under the Sea*) to Martel's magical-mystical tour in *Life of Pi*, the Euro-American Pacific is portrayed as site of fabled sublimity and figurative allure, uncanny enchantment, exotic grandeur, death, if not registered (amid castaways, whales, sharks, tigers, or giant cuttlefish looming up) at times as some Bali Hai call (as in some ever-playing musical *South Pacific*) to global adventure, tourist quest, and cultural otherness.[54]

Sea Slaves, Coral Reefs, Oceanic Cradles

"The Pacific beneath the pavements" can no longer mean recycled images of tourist-beach fantasy, as in *Blue Hawaii* (1961) and *Blue Crush* (2002), or surf festivals of oceanic conquest from Mavericks to the North Shore (as portrayed in Stacy Peralta's *Riding Giants* [2004]), for we are living not just in a precarious time of climate endangerment and class immiseration but also in a time when indigenous bodies of sweat and saltwater refuse to be commodified or exoticized across a decolonizing Oceania.[55] The Pacific beneath the pavements must, at the first ecological level, recall disappearing coral reefs and native islands being submerged, oceanic acidification, and thermal shifts amid the mounting North Pacific garbage gyres of transnational detritus between Japan and the United States. By the year 2050, with some eight million tons entering the oceans each year, there may be more waste plastic coursing in the oceans than fish.[56]

An ocean-conscious ecopoetics tied to forces of "oceanic becoming" and the scale of a "blue humanities" environmental approach will have to confront such matters to articulate the submarine forms, coastal cities, and oceanic sites affected by the hyper-capitalist world. As Mark Lynas summarizes this telos of waste production, "Our detritus gets everywhere, from the highest mountains to the deepest oceans: abandoned plastic bags drift ghostlike in the unfathomable depths, even kilometers beneath the floating Arctic ice cap. Wherever you look, this truth is there to behold: [for] pristine nature—Creation—has disappeared forever."[57] International search-and-rescue teams scouring the Indian Ocean off the west coast of Australia for signs of Malaysia Airlines Flight 370, as Barbara Demick observed, "discovered what oceanographers have been warning—that even the most far-flung stretches of ocean are full of garbage."[58] The ocean is not just blue but red, brown, and violet in its chromatic signals of injury and distress.

"Sea slaves" of contemporary oceanic labor across world seas are exploited on "ghost ships" remote from cities that eat the cheap fish or feed their pets at home; hence, this forced, alienated labor is unseen, or sublimated into what Allan Sekula has portrayed and documents in film and essay as the process of "forgetting the ocean."[59] The phobic warning placed on early-modern *mappamundi* maps to signify oceanic space as an unknown wilderness of eruptive monstrosity and world threat—"Hic Sunt Dracones" (Here be dragons)—may have to be reclaimed as planetary signal to register our own, displaced debris and supply chains of migrant sea labor.[60] "The unconscious was truly a *Mare Ignotum* [unknown ocean] when he first let himself into it," as Murray Stein has written of Carl Jung's quest to unlock psychic depths of oceanic subconsciousness, whereby "many of his most important intuitions originated in his experiences of the sublime, which came to him in dreams, visions, and active imagination [poesis]."[61] The flotsam and jetsam of far-flung world oceans, as they digest the everyday life of global capitalism and urban excess, constitute a feedback system of "blue" and "green" signs of planetary equilibrium and renewal but, at the same time, radiate a far more dangerously "prismatic ecology" of violet-black, white, gray, and red warnings.[62]

This oceanic waste and urbane unconsciousness is one effect of a global postmodernity that is full of uneven prosperity, disregard, and oblivion. At its American sublime origin, Walt Whitman's poetic persona, birthing self into voice in "Out of the Cradle Endlessly Rocking" by the Long Island shores of the Atlantic, can only hear the ocean murmuring some threnody of world mortality to self and world, "the low and delicious word death / And again death, death, death, death."[63] Whitman, like Jack Kerouac later, experienced the ocean as a "road" opening the self to the world in all its depths, risks, and perilous quest:

> Not I, nor anyone else can travel that road for you.
> You must travel it by yourself.
> It is not far. It is within reach.

Perhaps you have been on it since you were born and did not
     know.
Perhaps it is everywhere—on water and land.

Whatever natural sublimity the ocean retains as a road or way or nexus
of world transit linking land to planet and human beings to planetary
becoming, it is still commonly portrayed as a space that is inhuman, alien,
and deadly to inhabitable design as an element. The modernist poet Marianne Moore figured this antagonism felt by urban dwellers who confront
the ocean as an immeasurable, inhuman, alien, and deadly element:

It is human nature to stand in the middle of a thing,
but you cannot stand in the middle of this;
the sea has nothing to give but a well excavated grave.[64]

Bob Dylan's death-haunted album *Tempest* (2013) gets at the world-
shattering threat of *oceanic unmaking* through his dreamy narrative retelling of that world-capitalist disaster, the sinking of the RMS *Titanic* in
1912. In the song "Tempest," Dylan recounts this oceanic trauma in spectral stanzas as if to awaken contemporary citizens from their own, forget-
ful luxury spreading across world seas of postmodernity: "The watchman
he lay dreaming / Of all the things that can be / He dreamed the *Titanic*
was sinking / Into the deep blue sea." As if recalling the storm-tossed
Tempest and courtly shipwreck in Caliban's British Caribbean, Dylan's
Atlantic recalls engulfing seas "as a basic inhuman-ness, an alterity
that defines Shakespeare's ocean throughout his career."[65] Dylan's Beat-
influenced lyric, "A Hard Rain's A-Gonna Fall" (1962), had been prophetic
in its mourning for a nuclear-haunted planet, as in its blue-eyed-son lines
of prescient warning, such as, "I've been out in front of a dozen dead
oceans" and "where the pellets of poison are flooding their waters."

As Jeffrey Cohen has argued of the global confluences traversing the
Atlantic and Mediterranean oceans that led to the rise of New York City
as a littoral site of multilinguistic and cross-cultural conjunction and rupture, "Across spiraling planes (current, conveyance) as well as through
vertical engulfment (drowning, oblivion), the ocean is transport and
catastrophe."[66] The theorist Peter Sloterdijk warns in *Neither Sun nor
Death* that, despite our global circulating since the era of Christopher
Columbus and Ferdinand Magellan, "People born today do not develop
any oceanic consciousness—neither in the phobic nor the philobatic

[self-avoiding of dangerous objects] sense," leading Gaia into disaster zones of global ("spherical") forgetting and broken unity we face as planetary horizon.[67] While a material biogenetic object of planetary magnitude and geo-ecological concern, the ocean remains for Sloterdijk one of those "sublime imaginary constructs of wholeness" we cannot conjure into contemporary world pictures of the spherical globe, given the terrestrial-centric predispositions that often still hold.[68] The miraculous ocean goes on breaking up under oblivious urban unconcern, in processes of *deworlding*, despite the National's invocation in the recent pop song "Terrible Love," of ocean as the figure of nature's endurance amid broken romances: "It's a terrible love and I'm walking with spiders / It takes an ocean not to break, It takes an ocean not to break."

Global brands of pavement buildup are spreading across and along our oceanic planet as road, bridge, high rise, urban expansion, artificial island, and beachfront encroachments multiply. Many of these projects are being initiated by PRC capital as megaprojects cast, across land and sea, to the Global South. As Alvin Lim, a political scientist of the Asian Pacific region, summarizes this PRC globalization dynamism and transnational expansion: "China is currently in its 'new normal' of single-digit growth, so its government has been very busy creating business opportunities for its industries across the world. In 2013 President Xi announced the 'Silk Road Economic Belt' and the '21st Century Maritime Silk Road,' both of which involve transportation and energy megaprojects, especially things like high-speed rail [and transcontinental superhighways from the Atlantic to the Pacific across the Americas]. I suspect the Latin American projects will eventually be folded into the Maritime Silk Road."[69] A regime fascinated by such resource-extraction agendas of global development, from islands in the South China Seas to post–Bandung Africa, is wreaking environmental damage down the Mekong and sideways to Southeast Asia and enforcing the derecognition of the island of Taiwan as a democratic nation-state or even as a commercial entity.[70] Remote islands subject to regimes of capitalist infrastructure, flow, and risk are not that remote.

Bill Knott warned, in a rebuke to Beat writers' overdependence on fossil-fuel consumption to power their cross-country roads and fuel the "dharma bum" God quests across the massively industrializing planet: "Faster faster, never slow / on the road to ecocide."[71] Knott's prophecy of "ecocide" goes unheeded from Beijing to Washington, DC, even when Pope Francis I is trying in neo-Franciscan terms of environmental empathy and stewardship to awaken the capitalist world system to ecological

18

consequences and class imbalances, as in overviews such as his 2015 encyclical letter *Laudato Si*: "We have forgotten that we ourselves are dust of the earth (cf. Gen. 2:7); our very bodies are made up of her elements, we breathe her air and we receive life and refreshment from her waters."[72]

Oceanic Becoming

Despite long-conjured odds of oceanic forgetting and obliviousness on urban pavements, "visa-free travel all the way to Hawai'i" that had protested against the reign of the Berlin Wall in 1989 represents a freedom of movement, libidinal release, and mobile embodiment, called here a recurring dream of *oceanic becoming*. Along these lines of transformation and flight, "all the way to Hawai'i" signified some utopic beach spot at the ends of the oceanic Earth. That was what the German demonstrators had longed for, mobilized, called forth, and dreamed. A beach might grow near Checkpoint Charlie. Admittedly, this Euro-Hawai'i is remote, dreamy, and far-fetched, yet it abides as a kind of Pacific utopia (*no place*) that, nonetheless, a free citizen should be able to travel to—as if some belated distant isle of the blessed, as Friedrich Nietzsche evoked through his mentor rival, Richard Wagner. From the gold explorations of Spanish and Portuguese ships to the Great Merchant fortunes made by British, Dutch, and American ships traversing Asia and the Pacific, modern capitalism has linked its outreach to what the American philosopher George Santayana called "a poetic blue-water phase of commercial development" that may not so much have passed away as mutated in lyric power and utopic mode.[73]

For the stability of the ocean as site of late capitalist weather remains perilous, fraught with risk, threatened with loss and death at every turn, as early portrayed (for example) in Shakespeare's play *Merchant of Venice*, wherein Antonio's "mind is tossing on the ocean" as he awaits the return of far-flung commodity ships ("rich lading wrecked on the narrow seas") to the world maritime hub Venice: "He hath an argosy bound to Tripolis, another to the Indies . . . a third at Mexico, a fourth for England, and other investments he hath, squandered abroad" (act 1, scene 3), as Shylock mocks such transoceanic indebtedness to fortune. The sea—its currents, winds, waves, doldrums, storms—moving across overlapping circuits of exchange, from Shakespeare's Mediterranean and Atlantic to Édouard Glissant's "relational" Caribbean and Epeli Hau'ofa's *Pacific becoming Oceania*, is not just subject, topic, or background but a copresent trans-species agent of world-making and world-breaking force. Boat and coffin, as Melville suggests through Queequeg's Polynesian coffin and the *Pequod*'s deadly shipwreck in *Moby-Dick*, were often figured as interchangeable words in water-connected sites across Southeast Asia transiting the not-so-pacific Pacific. The Scottish fabulist Robert Louis Stevenson affirmed in *In the South Seas* (1896) that literature needed to connect islands and ocean across sites, such as Scotland to Samoa, as interwoven sites of conquest, settlement, commerce, creativity, and history: "I must learn to address readers from the uttermost parts of the sea." Oceanic becomes not just a content but a tactic and a mode of becoming, as I show in later chapters.

Lewis Lapham has documented (in an issue of *Lapham's Quarterly* devoted to iconography of "The Sea") recurring tropes, myths, and concepts of immensity and threat that tangle around the ocean. The poetics of the ocean long portrayed as *sublime*—as figured forth in writings from the Book of Jonah, epics of Homer, and aesthetics of Longinus through lyrical passages in Edmund Burke, Joseph Conrad, George Eliot, G. W. F. Hegel, Immanuel Kant, Melville, Eugene O'Neill, John Steinbeck, Simon Winchester, and more—stands corrected by the grimmer sciences of marine biology and ecology, tracing oceanscapes endangered by thermal alteration, environmental waste, fishing depletion, entropy, species death: sea waters "awash in non-biodegradable refuse—cathode-ray tubes, traffic cones, and polypropylene fishing nets."[74] This nonbiodegradable detritus, buried beneath pavements of overconsumption, disregard, and waste, is what Charles Moore has traversed and documented in the subtropical northern Pacific gyre as *Plastic Ocean*, which can "defeat even the most creative and voracious bacteria."[75] "At the very least, your days of eating Pacific Ocean fish are over," is how Gary Stamper puts this oceanic endangerment, tracking radioactive trace elements across the Pacific such as iodine, cesium, and strontium in the wake of the Fukushima nuclear meltdown, "20–30 times as high as the Hiroshima and Nagasaki nuclear bombings in 1945."[76]

Disappearing from Maps

Still I pondered these Pacific Ocean dynamics inside the House of World Culture in Berlin 2013, city and land where Hegel had lectured on the capital-mobilizing power of oceans across state forms and world history (*Weltgeschichte*) in 1822; where Johann Wolfgang von Goethe had initiated reflection on the contours and system of "world literature" (*Weltliteratur*) in 1827; and where Carl Schmitt had formulated the planetary "nomos" of earth and ocean as sublating path to global world-picture domination in 1942.[77] This Germanic Pacific was still being framed as some *void*, a long night, an invisible continent, remote, noncontiguous, libidinous, backward, erotic, exotic. In a survey of novels comprising the core forms and values of cultural capital and world making in Pascale Casanova's Paris-concentric "world republic of letters," Christian Thorne highlights "the near absence of concertedly transoceanic novels [that] is one of our literary history's oddest lacunae."[78] The only Pacific-based work mentioned by Casanova is Keri Hulme's *Bone People*, based less on Maori commitment to people, language, history, and place than is its postcolonial translation into her skewered world system as a work of modernist experimentation and transcultural borrowing from Europe.[79] It is not easy to remember the Nazi ocean of Schmitt or the will to American military-commercial of world oceans in Admiral Alfred Thayer Mahan, or even to picture the Anthropocene-threatened islands flung far across Pacific or Caribbean distances.[80]

The Pacific Ocean under these German pavements more commonly can become romanticized into ethnographic exoticism as in the still admired texts of Otto E. Ehlers, whose *Samoa: Pearl of the South Seas* (1894) was being reprinted as a rediscovered signal of the German (post)colonial interest in the Pacific. Or it can be romanticized via a Berlin-based artist such as Emil Nolde, whose modern expressivism drew on masks, totems, and the polytheism of the South Pacific to energize or mutate his art after Paul Gauguin–like journeys (1913–14) into wilder places and signifying systems of alter-culture, from Papua New Guinea and Pacific islands to Japan. At Luna Park in Berlin during the late 1920s, splendid baths were constructed with artificial waves, as if to bring the ocean rhythms of the seashore inside the domesticated modernity of the German city.[81] In Benjamin's between-the-wars Berlin, this feeling of urban modernity luxuriating in its own stability "begins on the asphalt, for the breadth of the pavements is princely."[82] The oceans become just urban scenery, not interlinked world space.

The German artist and writer Judith Schalansky produced works that conjugate far-flung islands in *Atlas der abelegenen Inseln* (*Atlas of Remote Islands*) by meditating on the un-homely fate of her native East Germany, cast from smallness and difference into utter oblivion on world maps: "Then I looked for my country: The German Democratic Republic. East Germans could not travel, only the Olympic Team were allowed beyond our borders. . . . It was pink and tiny as my smallest fingernail. . . . My love for atlases endured when a year later [1989] everything else changed: when it suddenly became possible to travel the world, and the country I was born in disappeared from the map."[83] Registering the long-standing European quest for planetary space and travel across oceans, Schalansky's *Atlas of Remote Islands* researches the far-flung fate of fifty islands across the world, naming twenty-seven islands in the Pacific, from those of Alexander Selkirk (alias Robinson Crusoe) on the Juan Fernandez Islands of Chile in 1704 to the fate of Iwo Jima, in the Volcano Islands of Japan, during World War II, tied later to the image of New York firemen in September 11, 2001, "the summit of Suribachi reborn on Ground Zero" as she portrays.[84]

All this would be subject to global disruption from above and from below, near and far—by Atlantic and Pacific wars and catastrophes on and beneath the pavements of death and destruction. Fritz Lang's *Metropolis* (1927) foreshadowed such urban ruination through the dystopian figuration of a massive subterranean flood erupting from beneath gleaming Weimar pavements and the built-up prosperity of 2026 to engulf the class-ridden city of world modernity. The "modern unrest" attendant on "spatial expansion resulting from Atlantic seafaring and the discovery of the New World" becomes aggravated with the rise of Japan, the United States, and China as financial and maritime powers across the Pacific. To invoke *You Must Change Your Life*, as Sloterdijk summarizes this disruptiveness as a "spherical" fate under maritime-*cum*-financial regimes of neoliberal globalization, "The primary fact of the Modern Age is not that the earth revolves around the sun, but that money flows around the earth."[85]

Back from Berlin, I started writing and researching *Oceanic Becoming: The Pacific beneath the Pavements*, to evoke utopic projections of beaches and Hawai'i / the Pacific from Paris '68 or Berlin '89 and move beyond these entrenched tropes ("the void Pacific," "the invisible continent," and so on) into dystopic oceans under the Anthropocene. Beyond this, the goal is to excavate an ecopoetics and politics of oceanic becoming in contexts of planetary threat. While it may seem that the well-being, safety, and prosperity of cities from the Battery of New York to the containerized ports of Oakland, Liverpool, and Los Angeles are connected to oceans, this relationship between an urban and a maritime global nexus is occluded, ignored, bypassed—all still but hidden "beneath the pavements." While working through this tactic of desublimated disclosure, I wanted to keep moving between city and region to world and globe, if only to channel Pacific-based obsessions scholarly and poetic that can make the prose informed about all of this. At times, we may have seemed to care more about the world entanglement of oceans, tropes, and regimes than many there did—or, at least, we wanted, as cultural critics and poets of the "American Pacific," to ponder the ocean more resonantly than as ethnographic romance, historical fable, or aesthetic foil. As Marc Shell elaborates, all but landlocked Germany has been long entangled in such oceanic connections, southward to the Mediterranean, or "*mare nostrum*" of Rome, and, all the more so in the wake of British maritime supremacy, northward to the Baltic and the North Sea, as sea routes to global power and imperial dominion. "Full like the sea is Germany's power" became the call to recuperated rule.[86]

A Forgotten Urban Embrace of Oceans

The Pacific Ocean here figures as *the sublime* in some double sense— miraculous and catastrophic, liberating and threatening—as if shuttling between utopian and dystopian affects, narratives, and images that would comprise or push toward the making of a "blue poetics." On the one hand, "The Ocean [Is] Full of Bowling Balls," to use J. D. Salinger's metaphor for the danger, risk, treachery, strangeness, death, trauma, irregularity, and murderous power of the ocean in an unpublished short story by that name set in contexts of a New England childhood innocence. In his wry poem, "Any Fool Can Get into an Ocean," Jack Spicer invokes an ironic rebuke to the ocean as inhuman space, estranged from any shared language with humanity: "Look at the sea otters bobbing wildly / Out in the middle of the poem." Spicer's solution reverts to a quasi-romantic one: "But it takes a Goddess / To get out of one [the Ocean]."[87] The ocean always recalls the Middle Passage and a history of catastrophes, as well as passages to transformation and redemption, as Claudia Rankine documents and enacts in *Citizen: An American Lyric* (2014): "No, it's a strange beach; each body is a strange beach, and if you let in the excess emotion you will recall the Atlantic Ocean breaking on our heads."[88]

"What the Sea Throws Up at Vlissingen" (1983), Ginsberg's poem cataloging industrial pollution in the North Sea, portrays the filth-flushing seas of the Netherlands—though, in these Anthropocene days, it might be set in the Pacific, as well:

> Plastic & cellophane, milk cartons & yogurt containers, blue
> & orange shopping bag nets
> Clementine peels, paper sacks, feathers & kelp, bricks &
> sticks,

Succulent green leaves & pine tips, waterbottles, plywood and
    tobacco pouches
Coffee jartops, milkbottle caps, rice bags, blue rope, an old
    brown shoe, an onion skin
Concrete chunks white pebbled, sea biscuits, detergent
    squeezers, bark and boards, a whisk-brush, a box top
Formula A Dismantling Spray-can, a whole small brown
    onion, a yellow cup
A boy with two canes walking the shore, a dead gull, a blue
    running shoe,
A shopping bag handle, lemon half, celery bunch, a cloth
    net—
Cork bottletop, grapefruit, rubber glove, wet firework tubes,
Masses of iron-brown-tinted seaweed along the high water
    mark near the sea wall,
A plastic car fender, green helmet broken in half, giant hemp
    rope knot, tree trunk stripped of bark,
A wooden stake, a bucket, myriad plastic bottles, pasta Zara
    pack,
A long gray plastic oildrum, bandage roll, glass bottle, tin can,
Christmas pine tree
A rusty iron pipe, me and my peepee.[89]

Oceans as threatened across the spaces and times of the Anthropocene,
Ginsberg's catalogue of ecological damage falls into an indifferent list of
near and far waste, human and nature thrown up into nature-culture *garbage*. Ginsberg's affect is one of not joy or cruel optimism but lyric defla-
tion, as sublime ocean turns into urbanscape of waste, filth, and oblivion
we need to contend with as planetary citizens of cities generating this
mess.[90] Maybe Boyan Slat, the twenty-three-year-old Dutch oceanic en-
trepreneur, is right: "To catch plastic [you have to] act like plastic," he
says, as he constructs (with crowdsourced funding) sixty plastic-catching
devices (concentrated in the northern Pacific to confront the Great Pacific Garbage Patch) in what is called operation Ocean Cleanup.[91] Still, as
Somerville has noted, while thinking through the Great Pacific Garbage
Patch both as US waste matter and as uncanny metaphor for Pacific diasporas and marine life along shifting scales of continental invisibility and
archipelagic interconnection, the "oceanic current [gyres] that produced

the garbage patch originally created another, more positive, archipelago—a concentration of plankton and other organisms" vital to life.[92]

In the chapters that follow, after a summarizing discussion of concepts and tactics crucial to "worlding," "migration," and the making of a "blue ecopoetics" in literature and poetics in key sites where such urban disclosure and world remaking takes place, I will zoom in on the Pacific Rim cities of San Francisco, Seoul, Kaohsiung, and Honolulu as oceanic-landed and global-local urban sites where such issues and tactics can be enacted. Crucial remains the island space of Taiwan, oceanic, green, and global in some exemplary sense, as I have experienced since 1995, when I was sponsored as a National Science Council research professor at major universities in the cities of Hsinchu, Kaohsiung, Taipei, and Taichung. Students wonderfully remarked, when I discussed Taiwan as an oceanic space caught amid Mainland China, Oceania, Japan, and the world, that the map was not so much a "taro" shape as that of a "whale" about to sail off into the great blue Pacific Ocean. For world oceans, as the poet-scholar-translator Kenneth Rexroth said of his own lively San Francisco, both connect and separate, a configuration he saw connecting the coastal city of San Francisco to the "Orient" of poets such as Hanshan, Li Po, and Rabindranath Tagore, among others: "One reason is simply that oceans, like the steppes, unite as well as separate. The West Coast is close to the Orient. It's the next thing out there. . . . San Francisco is an international city, and it has a living contact with the Orient."[93]

If dwelling in urban sites of organic farming and commitments to place that have spread across the Pacific Rim, oceanic citizens remain wary of becoming digital nomads in cities of the future. As the post-Beat African American experimental poet Tongo Eisen-Martin writes about the homeless-ridden streets of his home city of birth and death, San Francisco, and about a Pacific Ocean that at once surrounds the Bay Area city with a beckoning promise of blue immensity and locks its citizens into Blackness, service-job precarity, and abjection amid the urban confinement of joblessness, social failure, gentrifying displacement, police harassment, drug deaths, and crime on the pavements of urban blight in Hunter's Point and the Tenderloin:

> bet this ocean thinks it's an ocean
> but it's not
> it's sixth and mission.[94]

If, as the title of Eisen-Martin's first poetry collection, *someone's already dead*, urges of his Black urban polity, the ocean and its fabled bounty of nourishment, renewal, life source, interconnection, and promise may die in this forgotten urban embrace from San Francisco to the world. "I've been out in front of a dozen dead oceans," Bob Dylan warned in "A Hard Rain's A-Gonna Fall," his catalogue of ominous ruinations from the album *The Freewheeling Bob Dylan* (1963).

Still, it takes some critical and poetic doing to reveal this occluded urban-planetary nexus of transience, motion, and affect, even as the aim in urban-situated chapters is to enact potentials of transfiguration across a range of forms and tactics here called *worlding poesis*.[95] These tactics will be not just historical and geopolitical but ecopoetic, cosmopolitical, and experimental to world, meaning here, at a destructive extreme, *to deworld*, as well as to creatively *reworld*, this planetary saltwater element of peril and promise; intimate presence; and nourishment, if not (still in urban pavement contexts from San Francisco to Hong Kong and Berlin) obliviousness. I foreground these creative-destructive energies at the oceanic and riverine core of late capitalist weather along the Pacific Rim. Not just city sites but land farmed for centuries along coastal edges of the ocean, from the Atlantic and the Pacific to the Indian Ocean, are being lost to the rising salt tides and saline watersheds of the world seas. Urban living along the ocean inside precarious coastal cities of the Pacific Rim and the Global South (such disappearing archipelagic islands as Tuvalu) is coming to reflect what Ackbar Abbas (situated in a Hong Kong returning to a PRC state) has theorized as the uncanny "déjà disparu" (already disappeared) temporality of biopolitical life vanishing from local sites and global sounds within the Anthropocene.[96]

In *Oceanic Becoming: The Pacific beneath the Pavements*, I will build on the emergent regional framework called "Oceania" as a world-ocean frame of the interior Native Pacific peoples as theorized through an interconnected insular/archipelagic kind of thinking. I would also push "worlding poesis" (or world making as such) toward a utopic planetary way of projecting future space-time-world as an environmental horizon of comradely confederation in the interest of providing an ecopoetics to challenge the capitalist-driven telos of the Anthropocene—an endgame horizon that would abolish worlding tactics; concepts; and transfiguration of biopoetic belonging to the world at local, regional, and planetary scales.

Crossing sites within and along this coastal Pacific as linked to Oceania and "becoming oceanic," Oceania takes the lead in this transnational, transpacific, and transoceanic mode of ecological belonging to world oceans. I build on the Pacific-based work of writers and poet-scholars such as Epeli Hau'ofa of Tonga, Fiji, and Papua New Guinea; Teresia Teiawa of the Kiribati Islands and Honolulu (as well as Santa Cruz and Christchurch, New Zealand); and the Native Hawaiian poet Joseph Puna Balaz, who goes on forging his pidginized I ("pidgin eye") belonging to Oceania from the leeward coast of O'ahu and now from his diaspora in the US state of Ohio. I later invoke the musical and meaning-laden poetics of the last Hawaiian monarch, Queen Lili'uokalani, deposed by American settler forces in 1893, whose writings and songs encode her pleas for justice and her countermemory of nineteenth-century American history amid the "great powers" (*mana nui*) of the Pacific during her learned and cosmopolitan lifetime.

We are living the *deworlding* processes taking place across Asia Pacific, as elsewhere—meaning the dismantling of the ecological lifeworld threatened by multispecies endangerment, environmental destruction, extreme weather events, dismantled health plans and work regimes, resource plundering, global pandemics, and precariousness and cruelty taken as everyday norm. *Worlding*, at the core—as Donna Haraway, Anna Tsing, and Karen Barad, transdisciplinary colleagues at UC Santa Cruz in the environmental humanities and social sciences urge—would embody practices of *thickening* cultural-political and trans-species differences of resilient, care-driven, and sympathetic life survival on this planet. Instead of surrendering hope of change to this unmaking world, worlding can help to create other forms, possibilities, and values of world becoming, world making, and dwelling in the damaged world: aiming *to reworld the world* in some active, *gerundive* sense of remaking and healing local-planetary being that is not just beholden to capitalist temporality, prefabricated identity, or the regulated map grid of Mercatorian space as taken-for-granted horizon or urban life, diasporic crossing, and oceanic belonging.

I   WORLDING
      PACIFIC POESIS

ONE

Becoming Oceania

Ecopoetics across the Planetary Pacific Rim, or
"Walking on Water Wasn't Built in a Day"

As discussed in the introduction, the contemporary Pacific Ocean, extending from coastal California to Austronesia and Japan, as well as coastal and continental China, has become not only a crucial site of global capitalist dynamism and transpacific investment, but it also remains riddled with antagonisms of political, territorial, and commercial conflict any contemporary version of the "Pacific Rim" needs to conjure and contest. This contested imagining of the region has become necessary in an era when the interactive currents of globalization have led to the border-crossing deworlding emanating from these global capitalist antagonisms. This has resulted in the mutual demonization of the waning-superpower United States and the rising hegemon People's Republic of China (PRC) with its Chinese dream of a One Belt, One Road encircling the world oceans and crossing continental and national divides like some neoglobal Silk Road. In an environmental sense, we can at times all but *forget* this mighty Pacific Ocean that, in Herman Melville's phrasing from the whale oil-plundering and shipwreck modernity of *Moby-Dick*, "zones the whole world's bulk about it" into a grand water-and-air unity only whales and nuclearized submarines could now sensibly negotiate. We use this immense oceanic element of resource extraction and atmospheric renewal even as its myriad citizens from Shanghai to Long Beach and

Pusan to Auckland continue to build up and dwell within an urban life-world that depends for its continued material well-being on, from, and across this elemental ocean.

As in long-standing Romantic ambivalence, this ocean at once becomes troped as an alien presence of sublimity, an antagonistic threat, and yet as a wondrous source of curative powers, as in the deep-sea microbes that eat up the potent greenhouse gas, methane, that goes on contributing to global warming. This world ocean of alien obliviousness, climate misrecognition, and coastal real-estate enchantment can also flip into a site of industrial waste and plasticene remainder, as we recall from recurring oil spills to radioactive contaminants. These oceanic waters become filled with the heaviness of our military history and the damages of technological blunder. From the Bikini Atoll atomic-testing drills of Pacific Island displacement and environmental ruination during the Cold War struggle of the United States and Soviet Union for techno-supremacy to the latest nuclear disaster across the Pacific in coastal Fukushima, Japan, in March 2011, human agents propagating this capitalist modernity threaten not only the water commons and carbon-laden air of any locality but also the very Pacific Ocean weather and ocean currents as interactive planetary bioregion. The Japan-based tsunami reminded Pacific coastal dwellers from Sendai in eastern Japan, to the fishing harbor of Monterey and Santa Cruz in Northern California that the Pacific Rim is not just a discourse or a trope of economic unity. It is also an unstable, interconnected, fluctuating, stormy, and geologically interactive bioregion. Figurations of oceanic interconnection could, in effect, move us toward a vision of what I will denominate and evoke as the dynamic of "Becoming Oceania," projected as a necessary site of transpacific cross-border unity, as well as the possibility of transnational solidarity to deal with this regional, as well as planetary, crisis of oceanic and atmospheric belonging.

This forging of a *transpacific ecopoetics* (as articulated in Pacific-based writers such as Albert Wendt, Epeli Hau'ofa, Brandy Nalani McDougall, Craig Santos Perez, Gary Snyder, and Juliana Spahr, as well as the earlier beatitude-seeking writer of transatlantic oceanic crossing, Jack Kerouac) can help push urban-ocean-dwelling citizens toward figurations of what I call the "Pacific becoming Oceania." Such a world-remaking framework can help citizen-readers to link land and water by considering their abiding ties of planetary belonging and atmospheric connection, as well as their quasi-utopic dream of ecological confederation, world peace, and transracial solidarity across the Pacific. Between beachgoing, swimming, and surfing at the pleasure-zone extreme and shipwreck and industrial-waste disaster at the other, the ocean calls out for global reckoning through what Steve Mentz and others now call a "blue humanities" that goes beyond green ecological dreams of pastoral empathy and harmonious sentiment (between town and country) or the earthy heroic labors of the georgic (or agricultural farm labor–centered) genre.[1]

Vexed by aggravated animosities of post–Cold War history and forms of global antagonism, the global pandemic, and military crises of waning and rising superpowers, the South China Sea has become "Asia's Roiling Sea," as a *New York Times* editorial put it in the pre-COVID crisscrossing naval summer of August 2012: "The sea is not only an important trade route but is also rich in oil, natural gas, fishing and mineral resources. Nations are fighting over islands and even specks of rocks to stake their claims."[2] Pacific Rim sites, islands, and their modern nation-coded coasts and borders—from China, Taiwan, Korea, and Japan to Malaysia, Vietnam, Singapore, and the Philippines—are seen to be caught up in the "heavy waters" of military surveillance, resource extraction, biopiracy, and industrial waste, as Elizabeth DeLoughrey calls these threats of military-industrial intensification and risk.[3] Such conflicts of naval might and scientific supremacy can challenge the taken-for-granted telos of trans-oceanic globalization that (at least since the late 1970s) presumes that the regional infrastructure underpins and promotes "a [Pacific] Rim [unity] that is an imagining of transnational capital, [as] a co-prosperity sphere," as Christopher Leigh Connery, my colleague in "worlding project" cultural studies at the University of California, Santa Cruz has tracked over the course of four decades as a discourse, ideology, and mythic imaginary projected by the United States but also by Australia to the south and Japan to the north of these transoceanic routes.[4] Still, this bio-kinetic and weather-turbulent Pacific Ocean, which "zones the world's whole bulk around it [and] makes all coasts one bay to it" (to invoke Melville's coexisting trope of the sublime from *Moby-Dick*), remains uneasily amenable to such territorial demarcations, projecting as if presuming some brand of national, contractual, or marine sovereignty over the huge salty element known for what Mentz has called (in Northern Atlantic contexts) "shipwreck modernity."[5] From the Bikini Atoll atomic testing during the Cold War era down to the catastrophic nuclear disaster in Fukushima, Japan, in 2011, citizens and their militaries of late capitalist modernity (valuing security for resource extraction and exchange as well as 24/7 carbon-fueled economies) threaten not just the water, ground, and air of local dwelling in such sites, but also the interconnected Pacific as a planetary bioregion. This latest tsunami reminded Pacific Ocean dwellers from Sendai, in coastal eastern Japan, to Santa Cruz, along the coast of Northern California, as well as on the coasts of Oregon and Washington state (they long felt the impact of radioactive traces and tsunami debris washed up from Japan) that the Pacific Rim is not just a discourse or literary-political trope. It is also a geologically interactive bioregion threatened by industrial pollution and contaminated water, if not radioactive traces for years to come.

As argued here around the framework of alternative "worlding," we need a more compelling way to critically enframe this shared oceanic horizon as threat to modernity and promise to environmental futurity. We need to articulate another way to convert and conscript fellow citizens of this diverse Pacific Rim into a shared if wary figure of geo-poetic belonging, oceanic confederation, and ecological interest. This is one of the implicit aims here, as I evoke this oceanic consciousness-raising through the thick-descriptive conjuration of experimental poets and postwar writers of indigenous and cosmopolitan textuality whose work enacts forces of "oceanic becoming," suggesting, formally as well as in content, a different poetic and ethos of belonging (worlding as dwelling in) to this region and globe. We need this reworlding form rather than assuming the post-Hegelian will to geospatial dominion that has long reigned from Germany to England, and to the late imperializing United States and Japan of the Meiji era onward into industry, colonization, and war.

In an emergent environmental sense as well, we can blithely forget the ocean while dwelling in an urban lifeworld (in consumption-rich cities such as Shanghai, Honolulu, Kaohsiung, and San Francisco in the Pacific, or Berlin and London in the North Atlantic, for that matter) that depends for its precarious modern well-being *on, from, and across the ocean*. This *ocean commons*, if figured as a vital biospheric element necessary to sustaining life and planetary health, could help build up tactics and affects of ecological solidarity and modes of co-dwelling. To do so, the ocean would have to be framed in terms that can inspire an imagination of cobelonging, mutual interest, and care. To invoke one Pacific Island case of particular US and Japanese interests: Hawai'i—at least since the 1840s, when the US manifested its interest in securing Pearl Harbor as a strategic coaling station en route to markets of Asia, until its annexation in 1898—has long been entangled in US struggles to secure Pacific maritime space as some kind of extracontinental territory and element of modern dynamism. As the projectivist poet Charles Olson evoked Melville's post-frontier American drive for space as some kind of "oceanic deliverance" from shrinking space for settlement and resource extraction in *Call Me Ishmael* (1947), "Now in the Pacific. THE CARRIER. Trajectory. We [Americans] must go over space, or we wither."[6]

The pro-imperial diplomat Whitelaw Reid had summarized this US expansionist policy during its empire-contesting war with Spain (and implicitly with China and Japan, as well) in 1898 in both the Caribbean and the Philippines as oceanic archipelagos of national self-interest: "To extend now the authority of the United States over the great Philippine Archipelago is to fence in the China Sea and secure an almost equally commanding position on the other side of the Pacific. . . . Rightly used, it [military-stabilized expansion] enables the United States to convert the Pacific Ocean almost into an American lake."[7] Not just the Caribbean as American Mediterranean would become an American lake of contiguous interest but even noncontiguous sites in the Philippines and sites across the Pacific and Asia, from Honolulu to Guam, could become part of this American lake underwritten via a huge security-chain apparatus to this day. This move reeked of the Monroe Doctrine applied not just to the periphery of New Orleans and Miami but also to San Francisco, Oakland, Los Angeles, Long Beach, and Seattle.

As if activating its own version of a Monroe Doctrine in the region and shores closest to its own post–Middle Kingdom nation, China has adhered to a long historical claim to these sea-access routes to the Western Pacific and the Indian Ocean via the Malacca Strait, as Robert D. Kaplan argues in *Asia's Cauldron: The South China Sea and the End of a Stable Pacific*. He calls this oceanic and coastal space (geostrategically akin to the US Caribbean) a kind of "blue national soil" to the PRC navy.[8] Recalling world war history and the invocation of the Monroe Doctrine as a "great space" (*grossraum*) principle of foreign exclusion and imperial expansion, we might well be wary of its belated use in Asia and on the Pacific Rim.[9] Still, the larger geoterritorial context for this struggle presumes what is called "the competition for dominance in the Asia-Pacific region." Here, the United States has maintained its uneasy hegemony at least since the defeat of Japan and Germany in World War II, despite Richard Nixon and Mao Zedong's world-altering Shanghai communiqué of 1972 that affirmed, "Neither [side] should seek hegemony in the Asia-Pacific region and each is opposed to efforts by any other country or group of countries to establish such hegemony."[10]

"The Pacific is big enough for all of us," declared US Secretary of State Hillary Clinton—like a belated American Adam—at the 2012 Pacific Forum in Fiji. This did not satisfy China Rising, not to mention interior Pacific countries that had grown wary of neoliberal "Rimspeak" since the end of the Cold War.[11] US president Barack Obama's "Pacific Pivot" in 2011 toward increased US military presence has only aggravated the problem of definition, interest, and strategy, generating blowback and populist resistance in naval base sites such as Jeju Island in South Korea and Okinawa in Japan, as well as across the oceanic commons of so-called Moana Nui, from Hawai'i to Taiwan.[12] Along with twenty-one other nations and despite maritime tensions in the Pacific, China is for the first time participating in the US-led Rim of the Pacific (RIMPAC) naval and security exercises held in Hawai'i.[13] In short, caught between renativizing visions of "Moana Nui" or a resecuritizing RIMPAC, the Pacific remains riddled with antagonisms of political, territorial, and commercial conflict any version of "Pacific Rim" needs to conjure.

"Earth is a misnomer. The planet should be called Ocean," Ed DeLong has urged, registering his marine biologist's sensibility for the ocean as regenerative life fluid that constitutes some 90 percent of our biosphere.[14] We need to see ourselves as oceanic citizens as much as Earth dwellers, ocean-beholden peoples enmeshed in a tenuous, Gaia-like system. Here authors of oceanic ecopoetics, from Gary Snyder in the Pacific Northwest and Epeli Hau'ofa in Tonga to Craig Santos Perez and Brandy Nalani McDougall in Guam and Hawai'i as well as Juliana Spahr in California, can help disturb the environmental unconsciousness that often reigns. The ocean across Oceania can help to envision environmental stances and trans-territorial affects of planetary belonging.

As Rob Nixon has warned about the mounting imbalances of power, wealth, resources, risk, and vulnerability taking place across world oceans during neoliberalist regimes in this, our "geomorphic" Anthropocene: "Neoliberalism loves watery metaphors: the trickle-down effect, global flows, how a rising tide lifts all boats. But talk of a rising tide raises other specters: the coastal poor, who will never get storm-surge barriers; Pacific Islanders in the front lines of inundation; Arctic peoples, whose livelihoods are melting away—all of them exposed to the fallout from Anthropocene histories of carbon extraction and consumption in which they played virtually no part."[15]

Threatened with techno-human endangerment and systemic distortion of longer temporal and spatial weather patterns, the ocean calls for broader reckoning as species origin, resource, and horizon. Whales, dolphins, coral reefs, and marine microbes—from the era of Melville and Charles Darwin to Walt Disney and postcolonial works such as *Life of Pi* and *Cloud Atlas*—appeal for a sense of co-dwelling that would connect beings across various scales of lung-brain-blood-water-air linkage. "This connection of everyone with lungs" is how Spahr puts this imperiled wholeness in a collection by that name, tracing forms and forces of biopolitical relationality across militarized waters and polluted air that extend from the US Pacific Command at Pearl Harbor and Manhattan Island post-9/11 and post-Kyoto.[16]

Haunted by species extinction and disappearing islands and their primordial residents, Spahr's frame-shifting book *Well Then There Now* (2011) offers a work of experimental ecopoetics that is deeply oceanic in the way it situates Hawaiʻi and the United States not just in relation to Native Hawaiian struggles and multiculturalist movements in altered English, but also in relation to global forces of planetary depletion such as Arctic melting, species extinction, and resource extraction far and near. While Spahr challenges the first person plural "we" of Robert Frost's Manifest Destiny poem "The Gift Outright," from the Kennedy inauguration in 1960, she shows how US claims such as "this land was ours before we were the land's" fall apart in a native island space such as Hawaiʻi. Via global overlay of distant islands, she offers a more broadly oceanic vision of planetary and insular interconnection, indicting her own and "our" consumptive and polluting patterns, from her home state of Ohio to her residency as a university teacher and writer in Oʻahu and Northern California, back to Manhattan Island, where for a long time she was associated with the New York School of poets as radicalized formally by the language poetics of Charles Bernstein and his colleagues at the State University of New York, Buffalo. Here her focus becomes ecologically keen, linking near and far sites and consequences from melting glaciers and heating oceans: "They often lived on an island in the Pacific and they often lived on an island in the Atlantic. **Lake Chubsucker** They thought of these two residences of theirs as opposites although both were places of great economic privilege and resources, places that themselves consumed large amounts of resources and consumed more and more resources all the time. **Lake Sturgeon**."[17]

The "unnamed dragonfly species" and myriad freshwater and saltwater fish she lists, in broken catalogues, are invoked and collaged as endangered species, threatened with extinction by us-and-them binaries large and small, local and global ties, near and far spaces that are overlain in interconnected relation: "things of each possible relation hashing against one another," as she names this systemic process in another poem.[18] Land and ocean frames clash and split apart yet (as Spahr performs this action of conjunction in *Well Then There Now*) *hash together* in discrepant perspectives endangering the planet, thus calling out for a multirelational vision of commonality and care this work of experimental world-connecting poetics would enact.

These oceanic ties and flows in the interior Pacific, pulling back from the past and across into the future, also mark contemporary reindigenizing Hawaiian poetry, as represented across English-Hawaiian languages by Brandy Nalani McDougall in her carefully wrought poetry collection *Salt Wind / Ka Makani Pa'akai* (2008), whose "protagonist" binding family and self to history and place is not so much the human presence of mixed-race family members as it is the ocean wind full of salt resonances and Hawaiian remembrances, recalling modes of place-based and oceanic belonging. In her opening poem, "Po," she invokes a mode of primordial oceanic belonging via an updated, reframed translation of the Hawaiian epic poem *Kumulipo*, recalling a land-ocean space deeper than and prior to modern plantations, Pearl Harbor, and tourist history in Hawai'i:

> Before the land was tamed by history,
> the oceanside resorts and pineapple plantations,
> before the cane knife's rust, the dark time of sickness,
> the coming of cannons, the bitter waters drunk,
> before the metallic salt of blood, the rain emptied
> into rivers, the winds carved valleys and mountains,
> before the earth spurted fire, birthed islands.[19]

Re-worlding Hawai'i back into the salt winds and geo-energies of ancestral memory and postcolonial futurity, she opens readers to older, more oceanic modes of ecological cobelonging to sea and place. Her poetry is rich with language puns and place-name echoes, hidden by the Hawaiian language in the *kaona*, or layered meanings of figuration she has studied as a Native Hawaiian poetics of linguistic, as well as environmental and political importance. As McDougall elaborates in detail and scope of poetic, chant, and song reference, Native Hawaiians had connections to earth, sky, and ocean as living ties to primordial "ancestors" that serve as allies (cosmic gods become mentoring allies, or *kapuna*) to a life full of *pono* (balance), righteousness, the shared stewardship of nature, and *mana* (spiritual life force).[20] "Literature is the home of nonstandard space and time," as Wai Chee Dimock shrewdly remarks in *Through Other Continents*, and such works of literature can help bring these other, "nonstandard" spaces and times of the *Pacific-becoming-Oceania* into uncanny prefiguration.[21]

Book-length poems by the poet-scholar Craig Santos Perez also enact a related feat of repossessing Oceania and the Marianas, projecting a mode of world belonging in which Guam/Guahan can no longer be named (or forgotten as) just another "unincorporated territory" of the post-1898 imperial American Pacific. Resisting Guam's being called the "Pacific hub to Asia" in *from Unincorporated Territory [hacha]* and *from Unincorporated Territory [saina]*, or being referred to as "USS Guam." Santos Perez would resist the centuries-long Spanish and US "reduccion" process of "subduing, converting, and gathering natives through the establishment of missions and the stationing of soldiers to protect those missions." His poems (tied to transpacific tidelands as much as to the experimental writings of *Tinfish* in Honolulu and Bay Area open-field poetics) proliferate counter-namings and trace alternative routes and roots in a counter-geography of archipelagic belonging: not to an American Pacific (where Guam is a militarized security site) but to Oceania, thus challenging the call "to prove the ocean / was once a flag."[22]

In *Be Always Converting, Be Always Converted*, I aimed at decentering "American poetics" via elaborating long-standing transnational as well as local Pacific and archipelagic ties. In chapter 4, I focused on *the* crucial postcolonial figure of Pacific world reframing—namely, the Tongan writer and social scientist Epeli Hau'ofa (1939–2009), who turned away from social-science developmentalist scholarship and the reigning telos of capitalist hyperdevelopment that has gotten the modern ocean into the trouble it is in as a "Great Pacific Garbage Patch" ecoscape.[23] *Oceania* stands for Hau'ofa's New Pacific ecumene of counter-conversion, a strategic mode of refiguring this Pauline universality of address for Pacific Islanders for whom globalization discourse would hail them into market dependence, subaltern labor, and secular difference. Oceania—vast, watery, evocative, at core mysterious (like the earlier Papua New Guinea pidgin-vernacular term *wansolwara* for the Pacific as "one saltwater," meaning "one ocean, one people")—becomes a hugely consequential way of reframing and reworlding a new regional and global identity that has had wide impact and acceptance from the Pacific to the Caribbean as his work becomes aligned to the widely recognized poetics of oceanic/island relations in the postcolonial writer Édouard Glissant (1928–2011) from Martinique and the creole-writing Barbadian poet Kamau Brathwaite (1930–2020). As Hau'ofa shows, "South Seas," "Australasia," "South Pacific" (introduced via James Michener and Rodgers and Hammerstein during the postwar American hegemony in the Cold War ethnoscape of "militourism"), as well as "Pacific Basin" and "Pacific Islands," need to give way to *Oceania* as signifier of Pacific choice and plural coalition.[24]

Hau'ofa's paradigm-shattering figure for coalition building is called and figured forth as "Oceania," and the very poetic vagary of definition here becomes a resignifying form of watery unity by which Polynesian, Micronesian, and Melanesian—and any such colonial-imposed definitions of race or nationhood—could be shed as boundary lines, false confinements into smallness, irrelevance, and global dependence. *Oceania* was originally a French geographical term (*l'Océanie*) coined in 1831 by the French explorer Dumont d'Urville; it was also a still-used Roman Catholic signifier for the Pacific (starting from the appointment of a Bishop of Oceania to New Zealand in the 1840s). It has now become the name for one of eight planetary eco-zones on the Earth. Following Albert Wendt's call for a poetics and ecology of a "New Oceania," Hau'ofa framed his rebirth turn toward this resignified and refigured postcolonial Oceania along a transoceanic road leading from Damascus via Suva to Kona and Volcano on the Big Island.[25] This *Pacific-Becoming-Oceania* dynamic stands for a hope-generating turn away from insular models of smallness, lack, disconnection, or belatedness installed by colonial discourse and boundary lines in a different language and future.[26]

All but dematerialized into cyberspace *netizens* of urban space, urban techno-citizens dwell on the verge of "forgetting the [material] sea" as a site of cobelonging, resistance, and co-history, as the filmmaker Allan Sekula has documented in works such as *Fish Story* (1995) and *The Forgotten Space* (2011).[27] The ocean, in many sites across the globe, still remains the *unthought* as such in many ways, alien, antagonistic, liable to ruin beaches and coastal homes and to shipwreck oceans and drown swimmers in ways writers as diverse as Melville, Stephen Crane, Jack London, Virginia Woolf, Joseph Conrad, and John Cheever have depicted. We can all but forget this material-semiotic ocean even on a Pacific Rim that houses thirteen of the twenty largest container ports in the world, from Hong Kong to Long Beach.[28]

Living on landed edges, we trope the sea as *alien other*, as a quasi-biblical antagonist and site of the void, waste, or utter oblivion as in the

Jewish prophets Jonah or Noah, whose punishment for broken covenants was near-death by oceanic water. The ocean threatens to become some kind of blue-black abyss creating monsters from inhuman depths linking Job's Leviathan or Melville's *Moby-Dick* to *Aliens of the Deep* (2005).[29] The Pacific houses such deformed byproducts of our chemical transpacific waste in *The Host* (2006), the South Korean movie that links Seoul's Han River to the American Pacific, not to mention Guillermo del Toro's techno-monster film *Pacific Rim* (2013), with its bio-phobic vision of sea monsters (*Kaiju*, the Japanese word for Godzilla-like creatures) about to devour San Francisco, Tokyo, and humanity itself—*Pacific Rim*, the movie, generating an updated version of "Pacific Rim discourse" in a filmic register. "Pacific Rim" becomes less an ecological subject or political allegory than a market for circulating genre fantasies and threats from the future, despite a People's Liberation Army officer's prescient claim that "the decisive battle against the monsters was deliberately set in the South China Sea adjacent to Hong Kong" to provoke China and maintain US hegemony in the region.[30]

Pacific Rim recalls the ecological catastrophe of disappearing coral reefs and native islands being submerged, oceanic acidification, and thermal shifts amid the mounting North Pacific garbage gyres of transnational detritus between Japan and the United States. Any blue poetics of "oceanic becoming" needs to deal with such matters to register the submarine forms, coastal cities, and oceanic sites affected by the hypercapitalist world.[31] International search-and-rescue teams scouring vast reaches of the Indian Ocean off the western coast of Australia for missing signs of Malaysia Airlines Flight 370, as Barbara Demick observed, "discovered what oceanographers have been warning—that even the most far-flung stretches of ocean are full of garbage."[32] The flotsam and jetsam of the far-flung ocean digests the everyday life of global capitalism. An oceanic feedback system of "blue" and "green"—suggesting planetary equilibrium and ecological renewal—radiates a more dangerously "prismatic ecology" of violet-black, red, brown, and gray warning signs.[33]

As the German-based cultural critic and philosopher Peter Sloterdijk warns in *Neither Sun nor Death*, "People born today do not develop any oceanic consciousness—neither in the phobic nor the philobatic [self-avoiding of dangerous objects] sense," leading into disaster zones of global ("spherical") forgetting we face as shared planetary horizon.[34] While a biogenetic object of planetary magnitude and interconnection, the ocean remains one of those "sublime imaginary constructs of wholeness" we cannot conjure into the spherical globe, given our terrestrial predispositions that often still hold.[35] Our globalizations tend to remain dry and land-connected rather than wet, swampy, and dispersed ("wet globalization"), as Mentz has touched on in his study of saltwater figurations from containerized ships to swimming bodies in the Long Island Sound.[36]

We try to map the ocean with treaties and legal conventions by means of straight lines, contracts, and bounded spaces, in an uneasy dialectic of *mare liberum* (open sea) and *mare clausum* (closed sea) These projections of exclusive economic zones (EEZs) regulated by the 1982 United Nations Convention on the Law of the Sea (UNCLOS) fail to take lasting dominion. Such disputes over the boundary lines of oceanic positions is taking place in the ongoing dispute over uninhabited islands known as the Diaoyu in China and the Senkakus in Japan.[37] "China thinks of the South China Sea much as the U.S. thinks of the Caribbean [if not the entire Pacific as such]: as a blue-water extension of its mainland," to invoke Robert D. Kaplan, recalling the primacy of geography to international conflicts of land and sea amid collapsing distances and border-crossing euphoria along the global market.[38] US Navy Admiral Alfred Thayer Mahan's "mandated wide commons" of global commerce and military power defended as space of the US-secured high seas has shrunk in the wake of such United Nations–led territorial extensions of EEZs and resource scarcity.[39]

In his ethnoscientific study of microbial oceanography, *Alien Ocean* (2009), Stefan Helmreich elaborates a "dual [romantic-scientific] imaginary" bespeaking ambivalence toward planetary waters: the ocean has become both *trouble to us* (as with tsunamis and climate turbulence) and *in trouble from us* (as in our Pacific garbage patch or over-fishing). The ocean figures as elemental sublimity: an immensity at once threatening us and yet a healing "source of its own curative powers," as in these deep-sea microbes that eat up the potent greenhouse gas, methane, we overproduce globally.[40] Perhaps *the* Pacific sign of oceanic endangerment remains the contemporary global installation in the "ocean commons" created by waste and ecological unconsciousness on both sides of the Northern Pacific, from Tokyo to Los Angeles and back. This Great Pacific Garbage Patch is a gyre of plasticene detritus twice the size of Texas, weighing some one hundred million tons, that lies just below the ocean surface bounded by California, Hawai'i, and Japan. This slimy image of late capitalist sublimity is formed from throwaway bottles, chemical sludge, and polymers harmful to marine wildlife and planetary well-being.[41]

Such "ecopoetic" images of oceanic endangerment and waste could be multiplied: Tuvalu island disappearing due to global warming and rising tides; the US military buildup from Guam and Cheju Island to the Persian Gulf and Australia; the melting Arctic; mounting typhoons of Taiwan; the nuclear wastewaters of tsunami-stricken Japan. At the same time, John Luther Adams (whose music integrates sonic effects from his coastal environment in the Northern Pacific) has composed a symphony titled "Become Ocean," which won the Pulitzer Prize in 2014, as "a haunting orchestral work that suggests a relentless tidal surge, evoking thoughts of melting polar ice and rising sea levels."[42] "Life on this earth first emerged from the sea. As the polar ice melts and sea level rises, we humans find ourselves facing the prospect that, once again, we may quite literally become ocean" is how Adams explained the environmental context that led him to compose "Become Ocean," a work full of sonic echoes from oceanic waves, whales, and patterns drawn from planetary slow time.

Aesthetic figurations of oceanic becoming might help move urban citizens into a more transoceanic vision of solidarity, a sensibility of co-dwelling and shared atmospheric, viral, and climate peril. The Pacific might become what was classically framed as that "last world ocean" where the post-Greco Roman civilizational imperative—*translatio imperii maris*—falls apart into modes of rhizomatic becoming, multiplicity, and trans-species belonging. Building outward from this global commons spreading across Oceania, such a decentered imaginary of the Pacific Rim might move us toward "the promise" of the ocean as common space of altered belonging, ecological confederation, and planetary solidarity. The United States, fronting this nexus to Asia Pacific on many fronts, needs to overcome what in Singapore has been called its "self-identification [as] a missionary superpower" across this region of China Rising.[43] This is how Colin Channer reimagines affective ties to the oceanic presence around him, both as a mammal among other ocean-tied mammals on a reverberating planet and as a generational survival of the Middle Passage across the Atlantic, feeling "oblivion, solitude, silence," as well as (like a sperm whale from Melville's novel) sonic "quake roar" and the slippage of tectonic plates, in the poem "Spumante":[44]

> I know what it's like to be mammal
> filled with deepest ocean sounds:
> oblivion, solitude, stillness
> intermitted by quake roar,

tectonic slipping, lava fissures,
ship propellers drilling,
the human croons of whales.
There is slave in me, fat heritage.

Becoming oceanic here figures a threshold experience in some double sense: deeply miraculous and catastrophic, liberating and threatening, amazing yet horrifying, the ocean presence overwhelms land-centric categories of homeliness and bounded belonging, shuttling between utopic and dystopian affects, narratives, and images that would make up a "blue" ecopoetics. In the conjunctive hip-hop song by Mos Def (Yasiin Bey), "New World Water," the polluting effects of worldwide industrial capitalism and waste production of bottled water create a "lopsided" ecology of contamination that cannot be dispelled from drinking "New World Water":

> New World Water make the tide rise high. . . .
> The sun is sitting in the treetops, burnin' the woods
> And as the flames from the blaze get higher and higher
> They say, "Don't drink the water! we need it for the fire!"
> . . . .
> Flourocarbons and monoxide
> Push the water table lopside
> Used to be free, now it costs you a fee
> 'Cause oil tankers spill they load as they roam across the sea.[45]

In Mos Def's eco-Jeremianic 1999 chant, "New World Water" becomes a toxic, far-flung betrayal of whatever the "New World" stood for as redemptive promise in the Americas, across world oceans, Dystopia now rules the water. If the ocean has been figured in traumatic terms of war, disaster, shipwreck, monsoon, hurricane, and accident, the ocean abides as planetary origin: full of poetry, enchantment, and the magic blue waters of the humanities, as a turn back to oceanic writings of Jack Kerouac can help us to see even in this contemporary moment of what is coming to be called the perilous Anthropocene.

The first world *road* Kerouac ever dreamed of as a path across the globe and as an awakening *Künstlerroman* call to his lifelong vocation as writer was not American highway but world ocean. As he worked on a draft of his first novel in 1942, at twenty, he wrote, "Into this book,

'The Sea is My Brother' I shall weave all the passion and glory of living, its restlessness and peace, its fever and ennui, its mornings, noons and nights of desire, frustration, fear, triumph, and death."[46] The sea as global road Kerouac dreamed of—while still a student at Columbia University in New York City, where he had begun to meet Allen Ginsberg and William Burroughs, those life-hungry writers who would form the subterranean homesick Beats—was huge, world-encompassing, sweeping across the North Atlantic and the Mediterranean down to the Caribbean and around to the Pacific: "the old ports of Spain and Belfast, Glasgow, Manchester, Sidney, New Zealand; and Rio and Trinidad and Barbados and the Cape; and Panama and Honolulu and the far-flung Polynesians." *The Sea Is My Brother*, "the lost novel" published in 2011, survives as a rough sketch of those ever early energies, drives, dualities, dreams, flights, hints at the wildly oceanic style and scope that would give the world the body of work in prose, poetry, and prose poetry that Kerouac wrought.

Kerouac was writing this vow to his soulmate, Sebastian Sampas, in 1942, this "mad poet brother" from Lowell and fellow writer in a group they termed the Young Prometheans who were haunted by communism, war, brotherhood, love, restlessness, and literary ambition. Sampas would lose his life in the battle against Axis forces at Anzio in 1944, but Kerouac would remain faithful to the high writerly ambitions he first outlined in letters to Sebastian. He also would marry Sampas's sister, Stella, in 1966, as if to keep these ties to Lowell intact after the fame and abandonment hit. "Port debaucheries then back to sea" (38) went the rhythm of Ti Jean's lifelong deterritorializing flights from Lowell to New York, San Francisco, Mexico City, Tangier, Paris, and Florida and back to Lowell. "I hit the road [away from family and married life], bummed all over the U.S.A., finally took to shipping out" (38) says Wes, a thinly disguised, hard-drinking, quest-driven Kerouac. This lost novel shows Kerouac breaking not just with the academic life in the English Department at Columbia, but also with what Jack Spicer called "the English Department of the spirit, that great quagmire that lurks at the bottom of all of us," a kind of living death masquerading as middle-class normality and success.[47]

The Atlantic Ocean Kerouac would soon cross and labor on as a merchant marine on the ss *Dorchester* (a transport ship sunk by a German submarine in 1943) would provide this globally influential writer with his first huge metaphor for world brotherhood and incarnational longing in sentences that began to encompass the energy of body and soul fusing in Whitmanic rhythms of some libidinal-political en masse. Written in a drug-induced three-week burst in 1951 and not published until 1957, *On the Road* would bring this vision of crazed unity more fully into world-circulating recognition in the Beat idiom through bebop prose and the questing God-hungry figures of Sal Paradise and Dean Moriarty, energized by the postwar dream of a restless America opening up with "all that raw land that rolls in one unbelievable huge bulge over to the West Coast, and all that road going, all the people dreaming in the immensity of it."[48]

The ocean as a geospatial medium had proved to be Kerouac's first interstate highway system, and it led not across the continent to the wilder US "Left Coast" freedoms of California, but across global waterways to the world as horizon, pushing history beyond war and murderous rivalry into something more blessed and worthy of the human quest. *The Sea Is My Brother*, published with an array of other uncollected writings from those juvenilia years of self-formation that have also been gathered by Kerouac's hardworking literary estate in *Atop an Underwood* (1999), remains what Ginsberg has called "just a lot of reverie prose." Kerouac was a writer driven by such youthful dreams across the 1940s and 1950s and on until his death in 1969 at forty-seven. As Sam Sacks has noted of Kerouac in his review of *The Sea Is My Brother*, "He did not, ultimately, want to flee from his youth; he wanted to enlarge it and live in it perpetually."[49] "The ocean wild like an organ played" for Kerouac, who tired of that weary tune of war, slaughter, and death, as did Bob Dylan writing lyrics in his post-Beat wake.[50]

The sea was crossed as an ocean of war, risk, commerce, and death, but it also stood for oceanic consciousness and connection, the dream of cosmic unities and "shipping out" across imaginative frontiers—"[as] if California had stretched around the world back to New England" (*The Sea Is My Brother*, 35). Here were spaces of wonder, bliss, and terror Melville, Thomas Wolfe, Samuel Taylor Coleridge, Edgar Allan Poe, and Walt Whitman had opened up that still called to Kerouac—"for the future world of never-war" (27), as a humanist character rants in *The Sea Is My Brother*. Like Dylan, Kerouac is always early, always getting started and being reborn; always writing a drunken experiment that is utterly sober in range, reach, uncanny eruption. As Gilles Deleuze has noted of Kerouac's quest for transformation through writing and life, "Writing carries out the conjunction, the transmutation of fluxes, through which life escapes from the resentment of persons, societies, and reigns. Kerouac's phrases are as sober as a Japanese drawing, a pure line traced by an unsupported hand, which passes across ages and reigns. It would take a true alcoholic to attain that degree of sobriety."[51]

The trope-laden afterlife of this restless road to world becoming keeps going, as Kerouac has become "posthumously prolific," as David Barnett put it in the *Independent*: "The novella follows two characters, old-hand seaman Wesley Martin and Columbia professor Bill Everhart, who hook up and ship out for Greenland carrying war cargo—a journey that Kerouac himself undertook on the SS Dorchester. The plot is minimal, and in both style and construction the novel betrays Kerouac's immaturity as a writer."[52] While the execution leaves much to be desired, as reviewers have noted, the novel does show foundations Kerouac was laying for future work. "There are wonderful bursts of Kerouacian jazz-prose which break through the strictures of the conventional novel," Barnett notes, "and even then his ear for dialogue was sharp and naturalistic." Sam Sacks dismissed such Beat work for his *Wall Street Journal* readers: "Instead [of the will to finished novelistic form] there was Mexico, Burroughs and Benzedrine; 'On the Road' banged out in 20 days on a 120-foot scroll (dictated by the Holy Ghost, the writer told [Robert] Giroux, who rejected it); the veneration of spontaneous prose and the whole baggy, unending, self-aggrandizing iconography of the Beats."[53]

But in the precursor fiction to *The Sea Is My Brother*, and in later, more fully elaborated Pacific-based and global works such as *Big Sur* (novel) and *Lonesome Traveler* (travel book), the sea had loomed as a horizon of world infinitude and experimental movement to Kerouac as American mythopoetic maker of narrative aggrandizement, not just as the history-drenched space of torpedo-ridden danger, militarism, waste, servitude, darkness, lacklove, wanhope, and death: "To make the sea your own, to watch over it, to brood your very soul into it, to accept it and love it as though only it mattered and existed!" (47). This world-pitching and world-heaving sea of planetary grandeur and border-crossing quest called out to Kerouac as it had, earlier, to Conrad and Melville, as well as to the Harvard-defecting Richard Henry Dana, if not to the "Passage to India"–obsessed Whitman and the transpacific zendos of Gary Snyder: "Next to the smell of salt water," says this early model of the hero-mimicking Kerouac of romantic infatuation with journey and bohemian wanderlust across ocean and road, "I'll take the smell of the highway . . . Whitman's song of the open road, modern version" (58).

Kerouac's oceanic quest is at core not just material or military but figurative, driven by American writer models: "life at sea—[becoming] a

Thoreau before the mast" (76). As Martin's more learned sidekick, Everhart, hymns it about such drives to explore land and sea in *The Sea Is My Brother*, "the pioneer is free because he moves on and forgets to leave a trace. God!" (63). Kerouac's first novel over-signals its romantic quest with portmanteau character names such as "Everhart" and "Earthington," as the point of view shifts with unstable abandon, unsure of who is the hero. Still, an ordinary seaman is figured forth as if reborn into an oceanic communism shorn of class consciousness, for these fictive sailors of a war-torn planet had "entered into the Brotherhood of the Sea—these men considered the sea a great leveler, a united force, a master comrade brooding over their common loyalties" (113). The masculinist metaphors that circulate in *The Sea Is My Brother* must stand by now in stark, anachronistic contrast to the ecological mandate to planetary co-dwelling in *The Sea around Us*, in which Rachel Carson evoked the ocean as a maternal, caring, interconnected, transhuman element.[54]

Kerouac knew well the murky reach and depths of the military-industrial sea. His ear was marked and marred by it, as was his modernizing era. But he also invoked more joyous transformations of the ocean as world-transforming force that pointed toward another history and way of being: global space as calling out for the Buddhist compassion way of the "dharma bum." In his Pacific-based experimental poem "Sea" (appended to the ocean-haunted novel *Big Sur*), Kerouac invoked the myriad, multilingual, and globally crisscrossed voices (French, Chinese, Alaskan, Pomo, Spanish, English, Hawaiian, Japanese, Russian, and more) that go on lurking inside the ever lapping Pacific waves of this huge ocean, from sublime Big Sur over to Nagasaki and Red China, and even back to his ancestral Brittany origins in France:

> O the cities here below!
> The men with a thousand
> arms! the stanchions of
> their upward gaze! the
> coral of their poetry! the
> sea dragons tenderized, meat
> for fleshy fish——
> Navark, navark, the fishes
> of the Sea speak Breton——
> wash as soft as people's

dreams——We got peoples
in & out the shore, they call
it shore, sea call it
pish rip plosh——The
5 billion years since
earth we saw substantial
chan——Chinese are
the waves——the woods
are dreaming.[55]

Along these lines of oceanic interconnection awakening urban readers to world consciousness via the subterranean and transoceanic quest
for something different, Kerouac once advised his fellow Beat poet Bob
Kaufman—as they looked out on a New York City that was becoming
the Pax Americana Rome of global finance, world culture, and military
empire—to stay hopeful, vigilant, vision-hungry, upbeat, "Walking on
water wasn't built in a day."[56] "Walking on water" here means, as I trope
on Kerouac's prefigurative prose, the process of becoming oceanic as in
this Lowell mentor radiating planetary compassion and modes of cosmic belonging across space and time. Approached in this elemental way,
Kerouac can help us to learn "oceanic consciousness" even in urban sites
of 24/7 anxiety riddled with highways, urban pollution, and rush-hour
traffic jams: to learn how to flow and mix across solid lines and landed
borders; to learn how to coexist with the elements and other species; to
learn a mode of being and archipelagic belonging linking together both
waters and lands and what Brian Russell Roberts has called transnational
"border waters" beyond trans-American "borderlands."[57]

Such a poesis of reworlding pushes toward "alchemical" transformation of urban pavements to disrupt liberal complacency and a way of life
that runs on automatic commercial drives, expressed in this Situationist passage from Paris 1968 and its longings for world revolution. Such
writers and theorists conjoined in this reworlding process can help us
relearn how to form figures and affects of conjunctive beatitude, creativity both individual and collective, to generate modes of planetary
co-belonging and world-dwelling (worlding); to learn how to detoxify
new world waters on this one-world planet we share as dwelling place,
as abode of being and co-inhabiting across elements and species.
We must push toward such re-creative forces and modes (as poesis)

before it is too late and the Anthropocene arrives as everyday evening news filled with dry lightning strikes, recurring hurricanes, massive typhoons, wildfires, disappearing glaciers, and an ocean irredeemably altered from its long-standing equilibrium and atmospheric and riverine renewal.

TWO

Worlding Asia Pacific into Oceania

Concepts, Tactics, and Transfigurations
inside the Anthropocene

You don't have units plus relations. You just have relations. You
have worlding. The whole story is about *gerunds* — worlding, body-
ing, everything-ing. The layers are inherited from other layers, temporalities, scales of time and space, which don't nest neatly but have
oddly configured geometries. Nothing starts from scratch.... It's
about the thickness of worlding.

· DONNA HARAWAY, "A Giant Bumptious Litter"

Late Capitalist "Deworlding": Globalizing Is Not Worlding

The place-shattering practices, extractions, and displacements of world
capitalism go on distending the spatial and temporal sites, scales, and
resources of dwelling in the world, if not deforming the moral-cultural
ethos it takes for such diverse practices to survive on what Wai Chee
Dimock has called our "weak planet" of declining democracies, unsustainable ecological systems, unstable weather, climate threats, and species extinctions.[1] For if we are to activate tactics of resilient "worlding"
from here in coastal California across the transpacific to affiliated sites of
survivance in Seoul, Taipei, Kyoto, Honolulu, and elsewhere, "worlding

Asia" implies not so much Euro-derived theory applied as a *worlding of Asia* as it does situated practice of *worlding in Asia*.[2] It assumes the crucial difference between the worlding of Asia (*of* meaning tactics done to) and a worlding in Asia (*in* here meaning tactics enacted by the peoples in place), as Shiuhhuah Serena Chou in Taiwan, Soyoung Kim in South Korea, and I (in California) elaborate in the multisited collection of environmental works *Geo-spatiality in Asian and Oceanic Literature and Culture: Worlding Asia in the Anthropocene* (2022).

In shifting discursive contexts, Pacific Ocean formations have long designated—at least since the 1980s, if not earlier—regions called "Asia-Pacific" or "Asia/Pacific," the emergent "inter-Asia," or that enduring framework "Pacific Rim." Such re-formations of region, geography, nation, people, and place are not all that disconnected from earlier Atlantic, Indian, Mediterranean, or Arctic oceanic currents and trends of transformation situated within the Anthropocene as planetary epoch.[3] As I elaborate in this chapter, *worlding* should not be taken as just a gesture, theme, or tactic enacting global processes of late capitalism as normative telos within the modernity of the Anthropocene. Even an essay such as Haifa Saud Alfaisal's "Epistemic Reading and the Worlding of Postcolonialism" (2017) assumes that worlding is synonymous with the global-capitalist dynamics of world literature as a hegemonic system rooted in the metropolitan marketing of peripheries. Yet Alfaisal argues that we should attend to the "border gnosis" of *postcolonial* sites such as those in Qatar, the United Arab Emirates, China, and Singapore, to which his studies of the literary humanities would remain affiliated.[4] If the global is not the world as such, as scholars in the humanities from Pheng Cheah to Eric Hayot and others have delineated while advocating modern and postcolonial terms of difference, *worlding* should not be equated to the dynamic of neoliberal globalization as it still commonly is.[5]

Yet how can creative and research workers in literature, urban, or cultural studies actualize alter-temporalities or posit emergent spatialities "in the era of globalization," to invoke *The Worlding Project* (2007) collection that substantiated emergent differences between "globalization" and "worlding" as horizons of historical possibility, world-making, and world becoming?[6] This unmaking (what will be called here *deworlding*) of the world under the disruptive spread of globalization is what Jean-Luc Nancy means when he contends, in *The Creation of the World or Globalization*, that "the world has lost its capacity to 'form a world' [*faire monde*]: it seems only to have gained the capacity of proliferating . . . the 'unworld' [*immonde*]."[7] *Worlding*, as Donna Haraway urges to the contrary, means activating practices and tactics that would *thicken* and prolong differences of resilient life survival on a damaged planet.[8]

Instead of surrendering hopes of change to this unmaking world or unworld (*immonde*), worlding can help to create other forms, possibilities, terms, and values of world becoming, world making, and cultural-political dwelling in the world—in effect, aiming *to reworld the world* in some *gerundive* sense of remaking and local-planetary dwelling that is not just beholden to capitalist temporality, prefabricated identity, or the regulated map grid of Mercatorian spatiality as taken-for-granted horizon or urban life and oceanic belonging. As Édouard Glissant has suggested in his own world-making poetics of archipelagic belonging to the island-ocean-planet world of Martinique, as linked across the Caribbean Sea, this world echoes feedback (*les échos-monde*) at all points of contact and verges on the fruitful chaos of too plentiful relations (*le chaos-monde*) and ungraspability (opacity). It can also shift into the mobile totality of dwelling with and relating to (*la totalité-monde*).[9] To embrace this interconnected mode of archipelagic and transoceanic worlding is what Glissant posited (in cultural poetics) and enacted (in his literature) as a way of moving these island relations "from ethnopoetics to geopoetics to cosmopoetics."[10]

"Worlding Asia Pacific into Oceania," as I trope this world-making dynamic, would open up different ways of being with others, relating, and dwelling in and across this ocean-interconnected world, opening life-forms to what the cultural critics Pheng Cheah and Eben Kirksey have called "lived local temporalities" and ways of "being with" above or below the nation-state; reified regionality; species fundamentalism; opposing the world system of a carbon-fueled, resource-depleting, profit-driven capitalism running down the planet and depleting oceanic beings, as Ali Tabrizi's documentary *Seaspiracy* (2021) grimly portrays.[11] *Worlding* posited in this situated, multicultural, ethnographic, and oceanic sense needs to be embraced and thickened. Worlding can be articulated as what the anthropologist Aihwa Ong calls an "art of being global" that takes place without losing cultural-political differences that matter and preserving sites of refuge; as a practice that can have a worldly impact, in effect, activating world-making tactics within the spread of what has been called, in the Asia-based collection *Worlding Cities*, "planetary capitalism."[12]

Biopoetic forms of multispecies reworlding still occur under the transhuman, microbial, and ocean-entangled sign of what innovative biologist Donna J. Haraway calls "sympoiesis." Drawing on research from literary, cultural, urban, and ecological studies, this chapter conjures concepts, tactics, figurations, and warning signs of what "worlding" is or can or should mean living inside the "humus" (multiple earth forms) of humanity and working against Anthropocene "deworlding" compulsions. In a compressed formulation, *worlding* means the process of making anew or building up the lifeworld into differences that matter—*worlding* it. This worlding takes place as the production of lived resilient diversity and trans-species belonging and cocreating. To invoke Nancy on world-making as a process of becoming, "The unity of a world is nothing other than its diversity, and its diversity is, in turn, a diversity of worlds."[13]

To presume as normative that the contemporary period is tied to the telos of late capitalism (*late*, as in *waning*) is to assume that the worldview effect of historical transformation is laced with the systemic death wish of apocalypse (*late* meaning *over*).[14] It is as if this system of global-capitalist temporality and spatial displacement might collapse under the precarity of its own contradictions and eruptive riots and losses. This gets narrated as a telos that aggravates the ecological crisis to some kind of total planetary *endgame*. Along such lines, David Harvey calls this depleting of nature's resources and disrupting of ecological balance *the* "fatal contradiction" of neoliberal capitalism depleting nature undergoing states of eco-planetary crisis under the rise of the Anthropocene.[15] The worlding eco-crisis we face is more than just "another polluted river here or a catastrophic smog there" (255), as Harvey satirizes such carbon-fueled damages, even as he looks forward to the collapse of capitalism via some "humanist revolt against the inhumanity presupposed in the reduction of nature and human nature to the pure commodity form" (263). The wholeness of the world as everyday configuration was assumed by modern poet Wallace Stevens as a totality in itself: "The most beautiful thing in the world is, of course, the world itself."[16] Farewell to that stabilizing aesthetic nostalgia for oneness as such.

Still, any move toward universalizing the telos of late capitalism as a world-remaking force—what I call here the hegemony of "unworlding" effects—would absorb Global Chinese or Korean inter-Asian versions of situated locality and modes of global-factory capitalism into a melancholic temporal horizon of dismantlement, plunder, loss, and ruination. Given China's One Belt, One Road infrastructure project to englobe the world into a neo–Silk Road cast across lands and deserts as across world oceans, as touched on in chapter 1, the funeral rites of *lateness* and irreversibility shadow claims for such coprosperity or hopes for a geo-engineered ecological fix. An Anthropocene-haunted film such as Bong Joon-Ho's *Snowpiercer* (2013) captures this fatal impact, with its closed system of class warfare and technoscience catastrophe on train tracks of killer capitalist destruction, resource depletion, psychotic madness, drug use, and freezing planetary doom.[17] Anthropocene dread and doom undermine General Douglas MacArthur's triumphalist American postwar claim, "The history of the next thousand years will be written in the Pacific."[18]

We are living through an everyday *deworlding* across the Asia Pacific, meaning the dismantling of the ecological lifeworld as threatened by multispecies endangerment, environmental destruction, extreme weather events, dismantled health plans and work regimes, resource plundering, and a far-flung precariousness and cruelty taken as everyday norm. I am speaking here as a subject located within the devastating impact of Trumpism (*cum* Putinism) in the United States, an authoritarian trans-legal amplification of statist capitalism that embraces America-first profiteering, anti-multiculturalism, financial deregulation, hostility toward refugees, sporadic war, counterfactual spectacle, and fake news trolling to create social chaos, capitalist capitulation, nuclear trauma, and environmental indifference to global warming. The impact of this on the geo-military instability of Demilitarized Zone (DMZ) Koreas, North and South, and the peaceful coexistence of this "inter-Asian" region cannot be underestimated, even as the two post–Cold War Koreas move toward an "inter-Korea" of two (unequal) systems. McKenzie Wark phrases this huge Anthropocene-*cum*-denialist effect with post-Empire mockery while contending, in *Molecular Red*, that "the collapse of the Soviet system merely prefigures the collapse of the American one. While the ruins of the first are real and poignant, the ruins of the latter have not quite been apprehended for what they are."[19]

These *worlding* and *deworlding* practices of global capitalism are still connected across spatial and temporal scales such that the "inter-Asia" or Pacific Rim region—or, as Trump's regime calls it, as if to provoke Rising China as a neo-maritime power, a "Free and Open Indo-Pacific" region with headquarters at Pearl Harbor—are not so different from the transpacific cargo cults and transatlantic worlds constructed in tones of quasi-Jeremaic lament.[20] "The steel mills of Sheffield and Pittsburgh close down and air quality miraculously improves in the midst of unemployment, while the steel mills of China open up and contribute massively to the air pollution which reduces life expectancy there" (258), as Harvey puts this far-flung planetary impact as a creative-destructive system, from which these premonitions of "late" or "endgame" capitalism have been drawn. Asia itself, in all of its historical and cultural complexity, as the postcolonial Indian novelist Amitav Ghosh argues in *The Great Derangement* and shows in his India-based transoceanic novels *The Hungry Shore* and *The Glass Palace*, has played what Ghosh calls a "dual role as both victim and protagonist" in the Anthropocene.[21]

Asia generates this ecological crisis and impact, as Ghosh argues, powering up, in carbon-rife and population-dense urban sites from Burma and India to China, Singapore, Kaohsiung, and Hong Kong the "unthinkable" climate we face as seeming endgame blocking the imagination with hyper-objects and post-nature melancholia. We now confront the breakdown of affect, frame, and form.[22] In a nexus connecting Hong Kong and Mainland China, as June Wang documents, *shanzhai*, or "fake global cities," have sprung up at or near Shenzhen that specialize in commodity production (even in high-end artworks) using "high-profile citational practices that borrow symbolic images from established brands and/or advanced cities." In such sites, Wang argues, "Worlding city practices, or the art of being global [from below], entails an assemblage of urban initiatives that harness disparate ideas, logics, and techniques from various places" from both inside and far beyond China.[23] Such urban production dissolves any binary between authentic original and menacing reproduction along a transoceanic factory chain that has become as immune to climate critique as it is to aesthetic anxiety and indigenous claims.

Worlding, taken at some pragmatic level of reckoning posited against the rise of the Anthropocene as this Earth-planetary period is called, means that we have left the climate equilibrium of the Holocene abruptly behind. Worlding would create forms and values that serve to challenge the late capitalist *unworlding* of the everyday world, meaning that *unworld*, as Nancy calls it with disgust. As touched on, Nancy gestures toward the death-dealing *immonde* (117) or "glomus" (37) that goes on being delivered by the reign of globalization as a hegemonic world-order, the totality of the world-becoming-market as such.[24] To presume the anthropogenic form of such "precarious capitalism" as a total telos for the planet, however late or not yet modern, as Anna Tsing warns in *The Mushroom at the End of the World*, can block *transhuman* or *multispecies* forms of paying attention to those "patchy landscapes, multiple temporalities, and shifting assemblages of humans and nonhumans: the very stuff of collaborative survival."[25] Such forms of multispecies "reworlding," refuge, and cocreating are taking place under the sign of what Tsing's transdisciplinary colleague Donna Haraway calls a cross-cutting process of "sympoiesis." These are tactics of "making-with" (58) and yokes "for becoming-with" (125) that can create (as a mode of bio-cultural *poesis* or making) forms and stories of multi-species "ongoingness" (49) that push towards, figure forth, and enact survival amid the late-capitalist ruins of a damaged planet.[26] In effect, Tsing and Haraway engage in a *reworlding* battle against *deworlding* forces in the United States (as waning hegemonic force), if not across the planetary world riddled with delocalized, viscous, distended, and transboundary "hyperobjects" and the huge disruptions of the global pandemic altering borders, movements, and policies like never before.[27]

Zooming in on this compressed verb of crucial projective definition, *worlding* means building up, creating, and making lifeworlds—that is, actively *worlding* as act of poesis, oceanic and otherwise. Playing off *The Worlding Project*, my earlier post-Heidegger deformation and postcolonial definition of worlding, I had offered this summarizing definition: "WORLDING (v.)—*a historical process of taking care, setting limits, entering into, & making world-horizons come near & become local, situated, in/formed, cared for; instantiated as an uneven/incomplete material-cultural process of world-making and world-becoming.*"[28]

Working at frictional edges of anthropology, natural biology, and economy, Tsing frames these tactics and processes as those of a vast nexus of transhuman "worlding" taking place regeneratively in cross-border liminal or refuge spaces of toxic capitalism. For Tsing, *worlding*, and what she explicitly terms *reworlding*, means linking overlapping "multispecies" ontologies that might push beyond any human-ego presumed "Eurochronology" of androcentric authority crucial to modernity, as Hayot argues.[29] "Yet the modern human conceit is not the only plan for making worlds: we are surrounded by many world-making projects, human and not human," as Tsing affirms in her study of matsutake mushrooms that flourish in between the ruins and pours of capitalist development moving across forests, commodity chains, and auctions of a *transpacific* world-making via such exchange.[30] Tsing's entangled worlds presume multiple worlds working at far-flung scales of interaction, as she affirmed in *Words in Motion*, a related collection she coedited with the Japanese history scholar Carol Gluck.[31] Such a worlding focus, as in Tsing's study on "the world-building work of fungi," shows how these edible mushrooms manage to survive and flourish as refugees amid the death-dealing savaging and salvaging forces of planetary capitalism as everyday norm.[32]

Worlding becomes what, in a Pacific-based analysis of the "U.S. nu-clear empire" in the wake of the 2011 Fukushima techno-climate disas-ter, Yu-Fang Cho tracks as "an active, critical, and imaginative process that undoes the universalism and historical finality assumed by the term 'globalization,' and [projects] a future-oriented, emergent form of theo-rizing, activating, and writing about culture and geo-material politics."[33] Godzilla/Gojira resurfaces as the ambivalent figure of nuclear waste and postnuclear possibility for some Japanese-American eco-alliance and world peace. In *Reassembling Rubbish: Worlding Electronic Waste*, Josh Lepawsky has applied an ethos of "worlding" to trace the global-local ties and interconnected flows of electronic waste as produced, consumed, repaired, and recycled across "minescapes," "productionscapes," and "clickscapes" of electronics and what he calls the "discardscapes" being produced from North America to China and Africa and back.[34]

Lepawsky uses "worlding" as both a material *practice* and as research *ethos*: "As I use it," he writes, "'worlding' refers to two things: practices that bring together a jumble of not necessarily like things—people as citizens, consumers, and corporations; materials such as plastics, met-als, and glass; energy and information; sites and situations that become connected through the use and disposal of digital technologies—as if they cohered to form a common world[;] and at the same time the research practices that seek to map out and follow the actions of those doing the worlding."[35] Lepawsky keeps asking of this jumbled-together wasteful world: What is the right thing to do? This worlding and reworld-ing of waste matter back and forth across the Americas and the Pacific also marks and mars our contemporary Oceania, if unacknowledged as shared global ethos. What forms of worlding should we stand for? We need forms and tactics that allow for multiple forms of life flourishing and what Haraway has called "ongoinginess."[36]

Globalization functions as that "master term" we still commonly as-sume to reduce such a jumbled world to a "world picture" of ongoing cap-italist modernity, a worlding of plural worlds into one foreshortened and disenchanted world of endgame capitalist modernity across the Asia Pacific regional world. Yet "world," even in this transnational economic instance, remains an open-ended signifier of categorical, tactical, and tropological slipperiness that needs to be *disaggregated* into differing meanings—at least into what Hayot tracks as, first, "a set of systematic relations" imply-ing a self-organizing and self-containing totality; and, second, some kind of "ethical imperative" towards modes of more cosmopolitan co-existing and world-belonging; as well as, third, "a habitus and ground for human living" in some everyday anthropological sense of lived culture.[37] This latter sense is what Nancy (as well as Glissant) means by world as a totality of meaning for those who inhabit such a totality in motion. China has been pouring investment in infrastructure, aid, and debt structures into the island of Tonga as tied to the One Belt, One Road initiative and geopolitical linkages unheard-of in the postwar South Pacific. As the lodge owner Ola Koloi warned, these huge China loans should worry every Tongan. "I feel like I'll be Chinese soon," she said.[38]

Worlding as a material everyday practice has to contend with border-crossing forces and the transnational imagination amplified in an era of relentless mediated transmission, as David Trend tracks in *Worlding*, with its focus on "virtual" places and rematerializing practices and genres that impact the making of identity. As Trend argues in by now naturalized terms, "Worlding is about the process of change—about the things that influence the boundaries and rules that affect individuals, groups, and the dynamics [of identity-creating and place-making] that define their worlds."[39] Concerning sites of indigenous rearticulation, coalition, and emergence, Trend helpfully urges, "If a primordial consciousness guides us in the quest to sustain life on the planet, there is ample evidence to suggest that our newest technologies [of world making or worlding via digitalized interconnectivity] can be of help."[40]

In such global-local contexts of interactive communication and responsibility, it was in the second or more *ethical* sense of a lived imperative of quasi-bounded existence that Christopher Leigh Connery and I deployed "worlding" as a critical tool and advocated to enact a cultural-political-creative ethos of situated opposition in *The Worlding Project*, on which Trend's study builds in digitally mediated ways. Worlding becomes not so much a theme or subject as a tactic, practice, or projection (hence, *project*) of world making, projecting forth, as lifeworld becoming and building up, to use that Goethe-like word (*bildung*) for cultural formation. "Worlding" (as in "Afterword: Worlding as Future Tactic" to *The Worlding Project*'s troping on *welting)* means a mode of "building up a life-world palpably disclosing its lived-in modalities, boundaries, tactics, and historical processes" of survival and emergence.[41] Worlding stands for a range of such alter-globalization and situated practices of cultural-political localization being projected forward into and across the everyday world that is riddled and driven by mandates of capitalist dynamism and globalization. Worlding as such an ethos is built into the diffracting patterns of the quantum field theory world, as Karen Barad has elaborated. Barad phrases this far-ranging insight into the making of worlding as ethos: "Ethics is an integral part of the diffraction (ongoing differentiating) patterns of *worlding*, not a superimposing of human values onto the ontology of the world (as if 'fact' and 'value' were radically other). The very nature of matter entails an exposure to the Other."[42] Worlding in such "intra-acting" theorizing, in Barad's terms of tactile experimentation, "is about being in touch"—that is, "being responsible and responsive to the world's patternings and murmurings."[43]

From Deworlding to Reworlding

Worlding as projection and lifeworld experimentation are shadowed by residual forces and forms of *deworlding*; they also stand in emergent relation to those transformative life forces called (by Barad, Cheah, Haraway, Tsing, and others) *reworlding.* Worlding takes place amid contrary energies and practices of going and coming, deforming and forming, breaking down and building up lifeworlds. The former process (*deworlding*) suggests the enduring effects of empire, colonialism, and turbo-capitalism on other lifeworlds and threatened ways of being (as Ghosh portrays through transoceanic India environmental impacts in *The Great Derangement*). The latter process (*reworlding*) suggests, to the affirmative contrary, ways "to produce nuanced and situated tactics to surface alternative imaginings of the world we inhabit," as Chih-ming Wang shows in his multisited reading of Lawrence Chuah's diasporic novel *Gold by the Inch*.[44] By contrast, the "world" implied by globalization would aim to be the "most total totality of all," as Hayot puts this telos of modernity as what capitalism presumes to subtend for the planet as a world system—a totality of circulation and communication that has marked "world literature" since its origins in the cosmopolitical theories of Johann Wolfgang von Goethe and Karl Marx.[45]

Worlding as a critical practice enacts altered openings of time, space, and consciousness to other values and multiple modes of being, projections, and survival. Spatially, a worlded criticism seeks to form emergent connections to and articulations with region, place, area, and trans-species forms of being and dwelling on earth and ocean. Temporally, worlding allows for other modes of being in time than the instrumentalized "time is money" pragmatism of Benjamin Franklin and the New York City Wall Street and San Francisco Montgomery Street business-driven heirs to cash-value temporality of risk-taking and shorting as spread across the world. *The Worlding Project* aimed to undo the hold of what Radha Radhakrishnan has called "globalization as globality," meaning by that formulation an achieved global order in which counterworldings are seemingly over or irrelevant to this world becoming market system.[46]

Worlding, along these lines of collective creating, can remain mysteriously *gerundive*, akin to Charles Olson's breath-aligned poetic practices of postwar embodiment into "projective" verse; a world-forming verb more than pre-given noun, suggesting tactics to counter the norms or ruses of "globality achieved." By contrast, the world forming that takes place in the body's engaged movement through the world differs. "Walking shares with making and working that crucial element of engagement of the body and the mind with the world," as Rebecca Solnit contends in *Wanderlust*, "of knowing the world through the body and the body through the world."[47] As Stevens described his act of walking as remaking the world: "In my room, the world is beyond my understanding; / But when I walk I see that it consists of three or four / hills and a cloud."[48]

Worlding can be materialized in disparate ways to keep it multiple and emergent—a worlding that is not a fully identified critical object, as a projection creatively open to the present and future in ways that are not just those of the capitalist libido as "desiring production" system.[49] Shir Alon put this open-endedness of worlding in these terms: "The world is temporalized since it is never simply given or present, but is in a constant process of taking shape, a constant process of *worlding*."[50] Cheah's deconstructive *What Is a World?* marshals theory from Marx, as well as from Hannah Arendt, Jacques Derrida, G. W. F. Hegel, and Martin Heidegger, to track the telos of global capitalism and to enrich worlding tactics that would counter this taken-for-granted norm. The Anglophone postcolonial novel is interrogated to show how, as Alon summarizes Cheah's push beyond capitalist-domineered centers of Empire, "the sorry state of a globalized world [would] also carry a normative power to create and shape alternate worlds," as world "literature [becomes] an inexhaustible resource for world-making." Postcolonial literature plays a key role in these "counter-worlding" processes to those of world-system globalization; this happens in such literature by exposing the historical temporality of "unworlding," but this dynamic also seeks to grasp the "reworlding of the world" via forces tactics of decolonization—embracing not so much a *worldly* impulse (as Edward Said affirmed of de-Orientalizing critical practices) but a *worlding ethos* as such.[51]

This worlding of literature should not be collapsed into the worldly value of circulating inside, or battling for priority within, a unified market system of capitalist globalization, as Cheah sees happening among "world literature" theorists such as David Damrosch and the center-periphery world letters model of Pascale Casanova and Franco Moretti.[52] Market norm is portrayed as just one kind of world making, beholden to the circulation patterns and values of capitalist globalization. This mode has become posited as a model of worldliness and literary world making. Such global-driven modeling of world literature enacts "modes of world-making that ultimately make us worldless," as Cheah laments of this deworlding consequence of marketized world literature.[53] Worlding or world making, understood in the normative value Cheah wants to recover from cultural critics such as Goethe and Arendt as cosmopolitical vocation, means "a form of relating, belonging, or being-with" in the temporal sense felt as a "coming into being."[54] Worlding opens up different ways of being with others and being in the world in terms of connecting to other worlds and opening everyday life to other lived local temporalities" and ways of dwelling ("being with") above or below the nation-state or the bordered world system.[55] Worlding stays tied, in Cheah, to the Derridean figuration of a *justice to come*. Worlding is theorized as an "incalculable force" that can open up or, at times, oppose "progressive projects of world-making."[56] To invoke Alon's critique of this recursive reading of postcolonial time and narrative form, "The problem, rather, is the sense that Cheah reads postcolonial novels only to unearth, once again, the Derridian philosophical world-model."[57]

Worlding, as used in an Asian-situated multicultural and ethnographic sense tied to urban living across Asia Pacific, can become what Ananya Roy and Aihwa Ong, in their urban-sited collection *Worlding Cities*, call an "art of being global" that takes place without evacuating cultural-political differences that matter. Such urban modes of worlding can have a worldly impact, in effect activating world-making differences within and against the spread of "planetary capitalism."[58] Roy and Ong would affirm ways of being global that imply living (or *dwelling*, that most phenomenological of verbs) imaginatively in and across Asian-local and Pacific Rim urban sites, "at once heterogeneously particular and yet irreducibly global."[59] They see "the worlding city" as a lively, flexible, cutting-edge site of Asian capitalist invention, recognition, and experiment that cannot be reduced to "universal logics of capitalism or postcolonialism" or any diffusionist vision generated from Western urban centers such as New York, London, or Paris.[60]

"Worlding is the art of being global," Roy and Ong affirm, seeing Asian cities (such as Singapore, Vancouver, Dalian, Hong Kong, Dubai, Shenzhen, and Delhi) at a situated nexus of multitudinous construction and emergent solidarities, or modes of lived sociality, that they call (after Kuan-hsing Chen's *Asia as Method* [2010]) an "inter-Asian" referencing.[61] For this approach to *worlding the city*, "neoliberalism" is paradoxically all but adopted as "a set of maximizing rationalities," urban brands, and pragmatic cultural-political "outcomes."[62] In such urban sites along the Pacific Rim, worlding is thus theorized as everyday "art of being global." In effect, this means adjusting the urban lifeworld to the mandates, terms, forces, and forms of global capitalist real estate, as when Michael Goldman affirms "worlding through speculation and liquidation," transforming urban and rural lands, as what "it would take to turn Bangalore into a European-style world city."[63] Even "eco-city" projects such as the Songdo International Business District in South Korea and Clark Green City in the Philippines are said to assemble "the *spatialization* of *worlding* practices for countries to leave their mark on the world stage," to maximally fit the Pacific Rim locality into the mandates and forms of smart green globalization.[64] As Zack Lee describes such ecocities, "Worlding practices and visions intend to create global cities and global citizens but not everyone is able and is intended to take part."[65]

"Worlding aspiration," for Pacific Rim cities such as Hong Kong, Shanghai, and Singapore, means having a way to become hubs of financial speculation and world-city power; worlding here means modeling the future city on such a regulated capitalist grid of urban core empowerment and wealth.[66] As Chua Beng-Huat recounts this "1,000 Singapores" dream proposed at the 2010 Venice Biennale, worlding by a world-city becoming-Singapore model means "imagining the entire world population of 6.5 billion living and working in 1,000 Singapore-size islands, as a solution to global sustainability."[67] In another essay in *Worlding Cities*, Chad Haines depicts this world transformation taking place in Dubai; there, *worlding* means installing urban neoliberal dynamism so that "inherent in the chasing of the global dream is a transformation of self, through wealth, through status, and through the encoding of a rationalist, neoliberal self" adjusted to "the malls and workplaces where 'best practices' are instituted and regimes of surveillance are implemented," as in Beijing or Singapore.[68]

Worlding, as differentiated by Aihwa Ong in her introduction to the collection, means heterogeneous "projects and practices that instantiate some vision of the world in formation."[69] By this she means a kind of globalism circulated within and across urban sites, especially in sites such as Singapore, Hong Kong, and Shanghai taken as models. As Ong summarizes the dynamism that materializes the eco-urban future on the Pacific Rim, "Worlding practices are constitutive, spatializing, and signifying gestures that variously conjure up worlds beyond current conditions of urban living."[70] Such worlding practices inter-reference other Asian sites and inter-Asian practices, a mode akin to what Kuan-hsing Chen has called using "Asia as method" to interconnect, cross-reference, and deimperialize so as to build up "inter-Asia" local-to-local worlds.[71]

Worlding, in inter-Asian cross-referencing urban contexts, becomes a way to grasp and reinvent "the making and unmaking of the [totalizing] referent: Asia."[72] As Ananya Roy puts this, "Worlding is both an object of analysis and a method of deconstruction," as "inter-Asia" is invoked to destabilize and transform the making and unmaking of urban "Asia" into plural futures.[73] Other urban sites across burgeoning Asia see "counter-worlding tactics" such as traffic blockages or wildcat strikes, as what Roy calls "worlding from below."[74] "To ask where Asia ends and begins is thus to pay attention to the unstable space that is 'inter-Asia,' to trace the ways in which Asia travels, to make note of how urban experiments rely on the cautionary structure that is Asia," Roy affirms.[75] This gesture pushes toward worlding Asia along global capitalist trajectories somehow taken as different.

In "worlding multiculturalisms," Daniel P. S. Goh writes in another Asian-situated literary-cultural collection, *Worlding Multiculturalisms: The Politics of Inter-Asian Dwelling* (2015), the aim is to challenge the capitalist dialectic of state and society without "losing ourselves completely in the mediated circulations of global capital" or in the normative tactics of postcolonialism. Worlding means "those practices that infuse our arbitrary cultural lives with new things from cultural others in poetic ways to enable us to dwell and be at home with the complexity of the world."[76] Timely and multiple in urban focus and Pacific Rim interactivity, Goh's collection would activate across sites, from Singapore to Bangalore and Dubai, "interventions and interruptions of the dialectic that produce what we call 'worlding' moments of multiculturalism."[77] Worlding becomes here a mode of world transformation of the Asian global into dwell-able locality and sense of place.

Worlding Multiculturalisms tries to go beyond *The Worlding Project* and *Worlding Cities* by affirming Asian-sited multicultural tactics and dwelling places of urban and urbane worlding, which expressly aims "to grapple and understand emergent worlding practices and their emancipatory possibilities."[78] Describing the multicultural politics of urban advocacy, cultural maintenance, and ecological care that resulted in the preservation of Bukit Brown Municipal Cemetery in Singapore, which had been threatened with real estate displacement in its rapid urban growth, Terence Chong summarizes this *worlding Asia* approach: "The worlding of multiculturalism is an innovative process that sees the utilization of cultural capital towards the everyday objective of rootedness, belonging, and solidarity."[79] Even within the spread of casino-based capitalism in Macao and shopping-mall culture across urban spaces in Seoul, Vincent Ho and Sung Kyung Kim offer their essays as interventions wagering that "worlding multicultural" tactics can help to preserve Lusophone and Latinized difference—in the first case, "to express a multicultural Macanese cultural identity distinct from Chinese and Western civilizational imaginations"; and in the second case, to create public spaces of impure, mixed, or contradictory "dwelling" (that most Heideggerian of Germanic verbs) that allow for recognition of the competing claims of consumers, residents, officials, and sex workers in the Times Square mall in urban-regenerating Seoul.[80] *Worlding Multiculturalism* gestures toward tactics of worlding Asia as expressing modes of urban dwelling within normative globalization.

Using the political-economic signifier of *Asia Pacific* that the Chinese historian Arif Dirlik and I had called into question as an ideologically driven and economistic sublation of the Pacific Islands into Asia in *Asia/ Pacific as Space of Cultural Pacific*, I have called this chapter's elaboration of worlding dynamics, "Worlding Asia Pacific into Oceania."[81] Why *Oceania*? What does a broadly land-based, urban-driven, or mostly terrocentric inter-Asia vision of urban dwelling have to do with the uncanny saltwater, harbors, and far-flung islands of Oceania? Why that salt-watery world of the archipelagic Pacific? Why now? How can we move from the localist urban scale to a more planetary oceanic scale and still claim a bounded being or dwelling place? In "The Ocean in Us," the Tongan writer and social anthropologist Epeli Hau'ofa pushed toward articulating terms, values, mores, and forms of ecological solidarity under the figure he memorably called "Oceania." "And for a new Oceania to take hold," he urged, "it must have a solid dimension of commonality that we can perceive with our senses. Culture and nature are inseparable. The Oceania that I see is a creation of people in all walks of life."[82] Debating who belongs to this new emergent Oceania across the Pacific, Hau'ofa urged, "Oceania refers to a world of people connected to each other. [. . .] As far as I am concerned, anyone who has lived in our region and is committed to Oceania is an Oceanian."[83]

Belonging to Oceania becomes a matter of political and cultural commitment and conversion—not so much the *pathos* (suffering) or *logos* (argumentative necessity) as shared *ethos* of care. Oceania as a mode of dwelling means not only having a sense of history and cultivating caring attitudes and beliefs; it means also cultivating a sense of belonging to the planet and ocean as a bioregional horizon of dwelling, sharing, and care. Thinking with and beyond Hau'ofa's vision, Oceania can become for sites in Asia Pacific, as well as in other urban sites, (1) a worlding framework to help forge visions of ecological solidarity; and (2) a site of alternative modes of belonging inside Asia and the Pacific, reflecting Pacific and Asia linkages and knowledge formations. The oceanic imagination can prove helpful as a means of transforming social and regional practices in the making and shaping of a worlding ecopoetics.[84]

What has Asia (and, a fortiori, inter-Asia) to do with the ecological well-being of Oceania, its values, and its sites as spread across the Pacific to the Indian and Arabian oceans and back? By way of closing, I want to invoke two scenes from contemporary eco-documentaries, *Chasing Coral* (2017) and *Chasing Ice* (2012) directed by Jeff Orlowski.[85] In *Chasing Coral*, the bleaching and dying coral of Australia's Great Barrier Reef, as across broader oceanic spaces of the world, reflect intimate ties to the well-being and dietary needs of Asia, as documentary cameras across sites and regions are activated to show this impact.[86] In *Chasing Ice*, the biomorphic photographer Jason Balog uses time-lapse cameras from Greenland and Iceland to Alaska to track the huge melting glaciers and warming oceans of northern islands and waters that continue to impact and deform—or "unworld"—the oceans, archipelagos, coastal farmland, and deltas of India and China, as well as on the Korean peninsula and across the Pacific Islands.[87] These dying coral reefs and melting glaciers as transhuman ecosystems have destabilizing effects on "worlding Asia into Oceania," as the documentary shows, in a feat of trans-morphological and trans-urban-wilderness empathy.

As marine biologists and environmental scientists noted in a study of the Great Barrier Reef, oceanic research "confirms [that] coral reefs and acidic ocean water don't mix. Increasingly acidic oceans—caused by climate change—will harm coral reef growth over the next few decades if carbon dioxide emissions continue unchecked. . . . Besides their beauty, reefs shelter land from storms, and are also a habitat for myriads of species."[88] "Coral reefs are therefore the most biologically diverse ecosystems of the planet, and provide a number of ecosystem services that hundreds of millions of people rely on," Greg Torda of Australia's Center of Excellence for Coral Reef Studies has argued in Doyle Rice's overview. All of the world's oceanic coral reefs are at risk of bleaching from increasingly warm seawater: the frequency of severe coral bleaching events has increased nearly fivefold worldwide over the past four decades, along with acidification, Rice observes.[89]

Worlding—used as concept, tactic, trope, or warning sign—can help to motivate the preservation, linkage, and projection of endangered being and disrupted dwelling in such human and transhuman worlds. We are surviving precariously (if resiliently) across a not-so-lonely planet of coral and glaciers and much else endangered by the carbon-consumptive and resource-depleting system inside the Anthropocene. Worlding means making us aware of the ongoing bio-planetary dynamics of *deworlding* and (at the same time) calling out for other modes of multispecies *reworlding*. As the legal scholar Jebediah Purdy argues, the central question of living through the Anthropocene may come down to this problematic world-making dynamic and large-scale questioning. As Purdy summarizes the challenge of the Anthropocene, the "question [is] of what kind of world to make."[90] This chapter begins to suggest a worlding ethos tied to an oceanic poesis ("oceanic becoming") that is needed to help respond to and create a more responsive planetary sense of *worlding* and co-modes of sympoetic landed-oceanic becoming.

II WORLDING THE PACIFIC RIM

THREE

Toward a Blue Ecopoetics

Worlding the Asia Pacific Region into Figurations of Oceania at Monterey Bay

Those African persons in "Middle Passage" were literally suspended in the "oceanic," if we think of the latter in its Freudian orientation for undifferentiated identity: removed from the indigenous land and culture, and not-yet "American" either, these captive persons, without names that their captors would recognize, were in movement across the Atlantic, but they were nowhere at all.
· HORTENSE SPILLERS, "Mama's Baby, Papa's Maybe"

Counter-conversions toward "Oceania"

This chapter aims to invoke an emergent regional category and global-local vision of an Asia Pacific "ocean commons" coming to be called "Oceania" as an archipelagic interzone spread inside and across the Pacific Ocean to Asia and Australia.[1] The aim is to affiliate what others and I have called, at least since the end of World War II, the American Pacific. This oceanic turn also reflects the aim of transnationalizing archipelagic American studies by grasping "borderwaters" connection to modes of translocal solidarity, ecological alliance, and world belonging. Such an approach, as the "Archipelagoes/Oceans" special issue of the *Journal of*

Transnational American Studies avows, asks: "What if oceans and the islands they contain were perceived not as negative space around continents but as the positive space that defines postcontinental geography and throws transnational relationships into relief?"[2] At the core of this worlding-oceanizing tactic is the projection of an environmental-facing blue ecopoetics articulated with, and beyond, the Pacific and Asia imaginary of Epeli Hau'ofa's works in oceanic social theory and literature. In effect, the aim is to overcome nation-centric or naturalized bordered frameworks of a terrestrial Asia and Pacific identity that still reign from Auckland to Oakland and Hong Kong to Long Beach, despite the shared global climate crisis that comes from and returns to the warming and rising ocean waters.[3] A poetics of ecological reworlding, as discussed in chapter 2 on worlding Oceania, signifies a biopoetic building up of literary and cultural "poetics," which Jonathan Arac has defined in these history-laden terms: "Poetics as a term invites thinking about works of writing, or of culture more broadly, as the results of human making. What human beings have made may be studied closely in its emergence and its endurance in a historical world of human action and interaction."[4] This poesis is what I discussed in the previous chapter as an interdisciplinary basis for articulating ties to a trans-species *sympoiesis*.

Such a blue ecopoetics of world making would reframe sites such as California, Hawai'i, Taiwan, Fiji, Okinawa (in Japan), South Korea, and New Zealand and construct ties to an affiliated poetics of transregional worlding, bioregional sensibility, and translocal solidarity that has been long enacted in "transpacific dharma wanderers" and writers such as Albert Saijo, Jack Kerouac, Gary Snyder, Nanao Sakaki, Robert Sullivan, Caroline Sinavaiana, Sia Figiel, Susan Schultz, and Craig Santos Perez and others.[5] This oceanic turn cannot forget the traumas and transitions of a Black oceanic Middle Passage across the Atlantic and Caribbean, as Hortense Spillers reminds us in "Mama's Baby, Papa's Maybe" and "tidealectic" works such as Kamau Brathwaite's *Middle Passages* and M. NourbeSe Philip's *Zong!* portray, or the horrors of colonialism, war, exploitation, and racial division across the Pacific.[6] Maltroped and unpacific in cultural-political turbulence since the era of colonizing settlement, the Pacific (with its turbulent seas, active volcanoes, earthquake fault lines, depleting whales and dolphins, mounting typhoons and monsoons) calls out for ecological reframing, as does the postcolonial Atlantic.[7] "We return to the Great Ocean, which has been called the Pacific Ocean and the South Sea, two names which are both equally inappropriate," the German naturalist Adelbert von Chamisso noted of this global catachresis as early as 1821.[8]

This push toward articulating a broader ecopoetics of Oceania opens in this chapter by invoking three "local" urban images of the Pacific Ocean: *local* in this sense meaning the Monterey Bay in central coastal California, where I have lived and worked since I moved to Santa Cruz from Hawaiʻi in 2001. "Local" urban should have the multiscalar resonance of place-based specificity, agonistic struggle for social justice and native recognition, and globally entangled dialectics of world systems in which I have been immersed in Hawaiʻi and that I aimed to enact in *Reimagining the American Pacific* (2000), as well as later essays and collections.[9] The first image is that of human-dwarfing redwood sequoia trees spread across the mountains, watersheds, and hills of Santa Cruz. They have survived for centuries partly on fog that comes into Northern California from as far away as China and as near as Hawaiʻi but are threatened with drying pine needles and diminishing size by shifting thermal weather patterns across the Pacific. The second is the ocean floor at Monterey Bay, which is becoming a one-ton layer of human and military-industrial discards (such as artillery shells, fishing lines, bags, bottles, and plastic remainders), despite vigilant environmental efforts of forces and agencies that have agitated across the Bay Area since the 1960s to result in "Left-Green" and oceanic blue regulations. The third, apocalyptic, image forms a kind of global-capitalist postmodern artwork (so to speak): our global-waste installation aggregating across the transpacific ocean commons by a hyperconsumer-ist ethos and ecological unconsciousness across the Pacific Rim. As noted in chapter 1, the one hundred million-ton Great Pacific Garbage Patch lies just below ocean surfaces between California, Hawaiʻi, and Japan (Northern Pacific Gyre) and consists of discarded material harmful to marine wildlife.[10]

These three "ecopoetic" images could be multiplied across the region—Tuvalu Island in the South Pacific disappearing due to global warming and rising tides; the ongoing military buildup and damage from Guam and Okinawa; the naval port buildup at Jeju Island in South Korea—and added to the decimations across the Persian Gulf; the melting Pacific Arctic; mounting typhoons in Taiwan, Hong Kong, and the Philippines; and more. Such a transnational catalogue would be aimed at *converting* readers to a world-renewing vision of Oceania as a site of transpacific solidarity and ecological alert. In such perilous hazards, Steve Mentz has long elaborated this turn as the emergence and staying power of a multi-site and interdisciplinary "blue humanities."[11]

Oceania presumes a translocal regional formation of water-crossing linkage and multi-scalar reimagining built upon field-imaginary works in literature and social science by satirical fabulist Epeli Hauʻofa (1939–2009), to whom we remain indebted across the Pacific region for his capacious vision of islands, continents, archipelagos, and oceans in prolonged interconnection.[12] Hauʻofa's trans-islander regional frame can be shifted toward evoking a vision of alternative transnational belonging, ecological confederation, and transracial solidarity. Oceania, read as a performative speech act, enacts what Arjun Appadurai has called the "imagination as social practice," a public form of intellectual speech and cultural practice. Hauʻofa was spectacular at doing this in his own original, humorous, and visionary way, from Tonga and Australia to the Center of Oceania Art at the University of the South Pacific in Suva, Fiji.[13]

That loaded verb *convert* will give some readers *postcolonial* pause, hearing in this performative gesture quasi-imperialist echoes of latter-day "Orientalism" that Edward Said described in 1978 in these terms: "Yet the Orientalist makes it his work to be always converting the Orient from something into something else: he does this for himself, for the sake of his culture, in some cases for what he believes is the sake of the Oriental."[14] Said's formulation might forewarn the French Catholic philosopher Pierre Hadot or the American Jeremaic Bob Dylan away from any such Christianized acts of world conversion aimed at absorbing the other.[15] It is as if this conversion echoes in ironic reversal the Puritan tonalities of Thomas Shepard preaching to "praying Indians" in the Judaic-troped wilderness of colonial Massachusetts in the 1640s. When I gave a talk in 2010 on the conversion-troping study *Be Always Converting, Be Always Converted: An American Poetics* at the Academia Sinica in Taipei, a young Taiwanese scholar of literary-psychoanalytical persuasion remarked—rightly, to my mind—"Your title must mean 'be always perverting' for . . . the meanings and forms of conversion change in these Pacific and Asia contact zones [such as New Caledonia, Tonga, Hawaiʻi, and Taiwan]."[16]

That was what I had been aiming at: conversion taken in multiple fluid catholic (small *c*) senses rather than any monological sense of conversion in the Catholic (big *C*) sense, Roman or Eastern Orthodox, or any strictly magisterial church-based sense. Even in the classical Greco-Roman time frame, as Hadot has elaborated, the many classical and post-Christian meanings of *conversion* can range from *epistrophe* (return of self to an origin), as in the Stoics, to the more radical Pauline version of *metanoia* (mutation and rebirth) as a life-shattering transformation of flesh into spirit.

The strongest postcolonial portrait of contemporary conversion, in my study of literary and theo-poetic dynamics across the decolonizing Pacific, was drawn from Hau'ofa, for whom Christian conversion had become "reborn," translated, and transculturally mutated as a cultural form or world frame via his place-based and ocean-affiliated metamorphosis of post-contact indigenous belief into "counter-conversion" to Polynesian polytheism. Oceanic counter-conversion results for Hau'ofa in what I called a large-scale "regoding" of the Native Pacific Ocean into the presencing of Pele and Maui as cultural-political and environmental forces and a reworlding of islands into an oceanic nexus of spatial force. In effect, Hau'ofa enacted a turn away (*de-conversion*) from a neoliberal telos of capitalist hyperdevelopmentalism and its work ethic of production-consumption that has gotten the ocean states into patterns of dependence, not to mention the Great Pacific Garbage Patch ecoscape of waste.

As I described this counter-converting, in a chapter called "Writing Down the Lava Road from Damascus to Kona":

> Oceania as such invokes this New Pacific ecumene, for Hau'ofa, a strategic mode of refiguring this Pauline universality of address for Pacific Islanders for whom globalization discourse would hail them into market dependency, subaltern labor, and secular difference. Oceania, vast, watery, evocative, at core mysterious . . . becomes a strategic way of reframing and forming a critical regional identity. "South Seas," Australasia," "South Pacific" . . . , "Pacific Basin" and "Pacific Islands" all give way to Oceania as the self-identified signifier of Pacific choice.[17]

But in this postcolonial turn, counter-conversion is generated from the surrounding volcanic earth and sea on the Big Island in Hau'ofa's "Our Sea of Islands." The ecumene for Pacific Islander coalition building is called "Oceania." Its vagary of definition becomes a resignifying identity of unity through which Polynesian, Micronesian, and Melanesian, and any such colonial-imposed definitions of race or nationhood, could be sloughed off or cast away like dead boundary lines, false confinements into smallness, littleness, irrelevance, and global dependence.[18]

Hau'ofa framed his rebirth turn (speaking in Honolulu in 1993) toward Oceania—originally "l'Océanie," a French geographical term coined in 1831 by the explorer Dumont d'Urville, but this by now has become the name for one of eight planetary ecozones on the Earth—along a trope-laden road leading from Damascus and Suva to Kona and Volcano on the Big Island of Hawaiʻi where the volcanos remain so active and the island keeps enlarging into the deep blue ocean. Hauʻofa went on writing across two decades until his death in Fiji in 2009, driven by this hope-generating turn back to native gods, goddesses, myths, and art of Oceania and away from what he called "neo-dependency" globalization models of smallness, lack, or belatedness.

Ecumene is a worlding frame drawn and troped on from the prior Greco-Roman world, where it meant "the inhabited part of the Earth." World geographers still use it to signify populated sites on the planet. All the more so, *ecumene* and *ecumenical* have been used by religious formations since premodern eras of Roman Catholic dispensation to stand for unity and cooperation across divisions of faith. Hauʻofa is urging his own version of *ut unum sint* (let them become one again), as Norman O. Brown had recapitulated this drive to "unity" beyond modernist fractured differences in the systems of Pope John XXIII, as well as of Karl Marx and Sigmund Freud.[19] Hauʻofa gives this "ecumene" a more water-based and oceanic way of belonging to the contemporary world-enlarging globe: underspecified by borders and mores, "Oceania" becomes a transcontinental and trans-island framework whose center (echoing Ralph Waldo Emerson and Blaise Pascal on God) is everywhere inside the interior Pacific and whose circumference is not yet so fixed or certain, from its early use (which included Australia and New Zealand at the core) to Hauʻofa's later iterations (which at times can rather glaringly *exclude* Asia in toto) as discussed below.[20]

Oceania remains one of the nineteenth-century ecumenical categories of the Roman Catholic Church's global vision still used today. An early example of this would be *Early History of the Catholic Church in Oceania* (1888), by Jean Baptiste François Pompallier, Vicar Apostolic of Western Oceania and founder of the Catholic Church in Aotearoa in 1838.[21] A stained-glass window of Bishop Pompallier in the Catholic church at Lapaha, Tonga, Hauʻofa's familial homeland, calls him "1st Bishop of Central Oceania." Hauʻofa would have known such French and Roman Catholic usages that his later formulation of Oceania refigured for trans-indigenous and post-Christian circulation. Pompallier wrote his study in English, and Hauʻofa was implicitly mutating this prior theopoetic framework of Catholic Oceania into satirical reversal, as is his uncanny wont as a postcolonial fiction writer of the "Pacific Way."

Ecumene, as we can see in Gary Snyder's presciently ecopoetic usage, is related to the commonplace term *ecology*, the very "earth household" (as Snyder tropes on its etymology in his essay and poetry collection *Earth House Hold*) of water and land, economy and planet, that we have been pushing toward as a mode of more planetary belonging.[22] As Snyder writes in "Notes on Poetry as an Ecological Survival Tool" in 1969 (just before ecology would emerge as a global science across the 1970s), "Ecology: 'eco' (*oikos*) meaning 'house' (cf. 'ecumenical'): Housekeeping on Earth. Economics, which is merely the housekeeping of various social orders—taking out more than it puts back—must learn the rules of the greater [planetary] realm."[23] Hau'ofa's "Oceania" serves as postcolonial catholic (small c), boundary-shattering ecumene of Pacific Ocean belonging. *Oceania*, as envisioned in Hau'ofa, functions as a Pacific-based "God-term" (as Kenneth Burke called the terminological end term of such rhetorical persuasion), meaning a grandly inclusive synecdoche of part-whole belonging all but visionary in its boundary-crossing capaciousness and transracial newness shattering the codes.[24]

In social science and literary works as constructions of this interior "Oceania," in his collection *We Are the Ocean: Selected Works*, Hau'ofa shows that a process he calls "world enlarging" occurs across sites of pre- and postmodernity that he tracks in five ways:[25]

1 Via processes of material and semiotic exchange based on an ethos of "reciprocity"
2 Via seafaring and jet travel interconnecting sites
3 Via myths and visions of gods and peoples and sites
4 Via diasporic expansion and interconnection to "great cities" such as Auckland and the Bay Area in Oakland, San Mateo, Honolulu, and so on
5 Via this vision of "Oceania" as a counter-converting trope that turns the Pacific away from the telos of global domination and the ethos of disruptive developmentalism

In ecopoetic effect, Hau'ofa's Oceania serves as environmental framework of self/world and island/ocean/city interfusion to project a regional configuration at once ancestral and postmodern-global: spaces and times of premodern and modern connection by sea voyage and jet travel are relinked across modern colonial maps that render island peoples as small, disconnected, depleted, just some "invisible vast wilderness of islands" fly-specked on a tourist map, as Mark Twain put it in 1897.[26] Stories, images, art, dance, and legends perform and circulate a deep sense of

Pacific sublimity and spatial interconnection; long-woven networks of interconnected reciprocity prove crucial to this formation, as islands and oceans are interconnected across cities and countries; move, link, and flow across borders; and counter the late capitalist world from before, within, and after it. Hau'ofa, honored in life and death as so-called Chief of Oceania, in the view of the Papua New Guinea poet Steven Edmund Winduo, has opened a "school of thought," made a new beginning for Pacific cultural producers, and reframed Oceania as a "spirit of relatedness" on which we can build (as I am doing in this study) in other disciplines, paradigms, knowledge formations, and sites.[27] Oceania ties Pacific coastal cities such as Auckland, San Francisco, Oakland, Kaohsiung, and Seoul into one watery and environmental embrace, awakening the oceanic ties from oblivious burial beneath the commercial streets of these high-rise cities.

Blue Transfigurations of "Asia Pacific" Knowledge

Across altered spatialities and emergent knowledge formations, boundaries and study objects of areas and disciplines are being remade into frameworks like Hau'ofa's, analogous to those gathered in *Transnational American Studies* and related collections that alter the civilizational and national divides.[28] Regions/parts are becoming resituated into new wholes and linked to differing social and ecological energies. Pedagogies push toward articulating counterworlds and these forms of "new spatiality" that have emerged in literary, social-science, and cultural study. Reflecting modes of techno-interconnectivity and global mobility, globalization occurring from above and below national borderlines and forms, regions are becoming theorized as more open to fluid forms of relationality and interconnection—in frameworks such as "Oceania," "inter-Asia," "Americanicity," "Asia/Pacific," "the new Europe," and the "circum-Mediterranean"—than previous area studies had allowed.

  In the influential essay "Asia Pacific Studies in an Age of Global Modernity"—published in *Inter-Asia Cultural Studies*, a journal in which this interdisciplinary work in field reshaping and multisite region transformation of "Asia Pacific" has been involved since its founding in 1999, and with which my scholarly work has been affiliated—the postcolonial Marxist historian Arif Dirlik theorizes transformations of Asia Pacific field imaginaries as tied to shifting areas and regions since the late 1980s.

Dirlik and I were coeditors of the "Asia/Pacific as Space of Cultural Production" special issue of *boundary 2* that was published in 1995 as the expanded collection *Asia/Pacific as Space of Cultural Production*, which included Hau'ofa's frame-shattering "Our Sea of Islands." In the collection, we aggravated discrepancies among Cold War areas and Asian and Pacific studies while placing Asia and Pacific areas in rhizomatic, interlinked, transdisciplinary, and trans-spatial dialogue.[29]

As the introduction to *Asia/Pacific as Space of Cultural Production* declared, "The all-but-reified 'Asia-Pacific' formulated by market planners and military strategists is inadequate to describe or explain the fluid and multiple 'Asia/Pacific.' . . . The slash would signify linkage yet difference."[30] "Asia-Pacific," with the more solicitous hyphen of Asia-Pacific Economic Cooperation (APEC), weights the Pacific toward Asian sites as a source energizing and financing motions in labor, capital, and culture as we surveyed it.[31] "Asia-slash-Pacific" ("Asia/Pacific") can also mean opening up the Pacific to alternative formations, "as [this] Asia/Pacific region enacts the reconfigured space of nation-state deterritorialization, reinvention, struggle, and flight as power leaks out of the Cold War binary-machine."[32] Dirlik reflected on such disciplinary transformations via what he calls five "trends" that have arisen, across global academia, following on the crisis in area studies and the dismantling of Cold War rationales, or what Kuan-Hsing Chen calls a process of "de–Cold-Warization" across Asia and the Pacific.[33]

Dirlik's five trends are (1) civilizational studies; (2) the Asianization of Asian studies; (3) indigenous studies; (4) oceanic studies; and (5) diasporic studies.[34] While he sees the first three as largely "continuous with [area studies' and nation-based formations] in terms of fundamental spatial assumptions" of borders and fields via nations, he goes on to discuss oceanic and diasporic studies as representing "novel spatialities" that have arisen to challenge and assert alternatives to area- or nation-based models that had solidified during and after World War II (161). What Dirlik calls "the Asianization of Asian studies . . . directed against the hegemony of Eurocentric knowledge [of the area], especially United States domination of scholarship," and measured against an opposing turn to "insiders' views of Asian problems" and theories (164), might well be framed (in interior Pacific contexts) as the "Pacific indigenization of Pacific studies" or the worlding of Asia Pacific. This overlaps with what he separates into trend three: "indigenous studies," wherein he draws on the work of Vilsoni Hereniko and others (162–63). This turn would be directed against Australian, the American Pacific, and European claims to priority of a knowledge-power perspective and fit into "critical oceanic studies."

"Oceanic studies," while related to Pacific Rim studies of transpacific capitalism in Asia Pacific and other world-ocean sites, can serve discrepant global, national, and local interests. As Dirlik phrases this dialectic, "Oceans may represent projections of place-based indigenous ideals into space, as they do for Epeli Hau'ofa, or they may be used in service of an APEC version of space in the service of capital and [transnationalizing] states" (167). Gazing across the Pacific to Japanese Zen and Chinese poetry, as well as to communist movements in Russia and China, Kenneth Rexroth noted this possibility of linkage and difference in 1971, in an interview about San Francisco as connected to Asia Pacific by the ocean: "Oceans, like steppes, unite as well as separate. The West Coast is close to the Orient. It's the next thing out there. . . . [San Francisco] is an international city and it has living contact with the Orient."[35]

Dirlik's five trends—what we can call, after Michel Foucault, "power/knowledge" shifts in disciplinary paradigms—are summarized as "civilizational revivals" that bring about "the new attention to oceans, controversies over inside/outside forms of knowledge, diasporic motions and indigenous movements" (167). Such revivals all point to transformations in the making of Asia Pacific and the Pacific region as such, the rise of Oceania as well as the marketized allure of Global China. These transformations need to be implemented with critical caution, lest we return (say) not to Oceania but to the "Glorious Pacific Way" of the neocolonizing "Forge Foundation," as Hau'ofa warned through his faux-indigenizing character Ole Pacifikiwei in a short story by that name, who sells out Pacific oral histories and goes on to become a "first-rate, expert beggar" for global capital and island-sublating globalization that does away with island texts and ocean worlds.[36]

Asia Sublates and Unmakes the Pacific

Unmaking these *deworlding* colonial modes with raucous satire, Hau'ofa's fictional writing takes place in the post-British "many Englishes" of Polynesia, sometimes called by Juliana Spahr and others "alter Englishes," which are often creolized and pidginized in form. So pidgin artists (such as Joe Balaz and Wayne Westlake) can and do figure in the remaking of this *reworlded Oceania*, resisting the reign of what Hau'ofa mocked as

"World Bank English."[37] Hauʻofa's works at times demonize, mock, or exclude contemporary Asians from having affirmative claims to, or roles in, the construction of this alternative Oceania of ecological belonging and cultural-political resistance. In his satirical novella *Tales of the Tikongs*, Japanese corporate forces are linked to Pacific Rim operators from Australia and New Zealand.[38] "The Pacific Way belongs to regional Elites" (46), building cars too small for hefty Tikong people (12) and a tuna cannery that ends in disarray in "The Tower of Babel" (19–21). The gaming parlors of Taipei and sex shops of Tokyo and Sydney conspire to turn Tiko into "the South Pacific Haven for Gambling and Prostitution" (81). Pacific developers such as Ole turn to "regional [money] laundry centers" in Bangkok, Kuala Lumpur, Manila, Suva, and Moumea to learn how to do this sublimated mode of Asia Pacific exploitation (92). In a carnivalesque vision of Asians in the Pacific, the novella *Kisses in the Nederends* centers on New Age modes of duping, tranquillizing, and conning the indigenous Pacific body of Oilei Bomboki via that sage, yogi, and con man of Asian capitalist yoga and mind-numbing libidinal love, Babu Vivekanand.[39]

This subaltern interior Pacific is staged against the more instrumentalizing and expansionist "Asia" as figured forth in Hauʻofa's fictional works as described, as well as in his Pacific-based turn against Asia in the essay "The Ocean in Us" as sites *not* yet part of this emergent Oceania, suggesting instead the cognitive-mapping of US/Asia commercial forces aligned in hegemony over an interior Pacific. In this imaginary of Asia Pacific, Asia sublates the Pacific into neoimperialist dynamics, with the sea becoming capital's element of expansion across East and West, the Pacific as transnational space of sea power and ocean commerce, "liquid capital" coming and going from Asia, as Christopher Connery termed this telos of Pacific Rim forces.[40]

In "The Ocean in Us," an essay based, as noted, on the keynote Oceania lecture Hauʻofa delivered at the University of the South Pacific, Suva, in 1997 (which was published in *Dreadlocks in Oceania* in 1997 and served as the ecological keynote address in *We Are the Ocean* in 2008), Hauʻofa pushes toward forms of ecological solidarity. "And for a new Oceania to take hold," he urges, "it must have a solid dimension of commonality that we can perceive with our senses. Culture and nature are inseparable. The Oceania that I see is a creation of people in all walks of life."[41] Earlier, debating who belongs to this new Oceania, Hauʻofa urges in the same essay, "Oceania refers to a world of people connected to each other. . . . As far as I am concerned, anyone who has lived in our region and is committed to Oceania is an Oceanian."[42] Belonging to Oceania becomes a matter of ethical-political and cultural commitment. *Oceania* means not only having a sense of history and cultivating a set of attitudes and beliefs; it means a sense of belonging to the Earth and ocean as a bioregional horizon of care.

Later in the essay, after positing this capacious ethos of oceanic consciousness, Hauʻofa goes on to claim that in this Oceania, "Asian mainland influences were largely absent in the modern era," and that more specifically speaking, "Pacific Ocean islands, from Japan through the Philippines and Indonesia, which are adjacent to the Asian mainland, do not have oceanic cultures and therefore not part of Oceania."[43] But how can Asians, with sites such as Okinawa, Cheju Island, and Taiwan, be excluded by history, tradition, ethos, and territorial site from belonging to this new Oceania? Questions haunt sites in the Pacific and Asia: Can Asia become (then or now) part of Oceania? Cannot Oceania become the basis of a broader environmental collation? How can Oceania as such alter hegemonic "Pacific Rim" and "Asia Pacific" frameworks?[44]

This view of Asia as *sublating* the interior or subaltern Pacific or positioned outside of Oceania as ethnopolitical ecoscape is understandable in historical terms and not uncommon in a range of works and genres. In her poem "Amnesia," Teresia Teaiwa captures such a Pacific-evacuating Asia Pacific when she writes: "They're after American Pie in the East and some kind of Zen in the West. . . . So it's easy to forget that there's life and love and learning between Asia and America."[45] In *The Shark That Ate the Sun* (1992), the Pacific modernist painter-novelist John Pule sees this Asian base-linked Pacific turn into an "American Lake" for the US Navy, linking "ships in Samoa / Hawaiʻi, Taiwan, the Philippines, / Belau, Kwajelein, Truk / The Marianas, the Carolines"—a security chain across Oceania in which "the dead [as at the Bikini Atoll] are louder in protest than the living."[46] Techno-reformations of Asian and Pacific spaces and bodies are figured, as well, in theatrical performance pieces of techno-Orientalism such as Karen Tei Yamashita's *Anime Wong: A CyberAsian Odyssey*, as are transpacific Maoist links forged between China and Asian American struggles for Bay Area multicultural decolonization in *I Hotel*, her novel of transpacific and trans-American revolutionary heritage.[47]

In "Shrinking the Pacific," the Japanese American poet Lawson Inada imagines a shrunken, water-displaced Pacific Rim across which global travelers can "take the gleaming bridge / and bop into and around Hokkaido for lunch. // Maybe stay the night, or come back to Oregon, / which, by now is full of Hokkaido tourists" (or neighbors; it's hard to tell anymore in this unified Asia Pacific).[48] Joe Balaz, in a poem published in the web journal *Otoliths*, depicts the Hawaiian watershed at Waikiki that has become a shopping-mall carnival of fake cultures and clownish versions of indigeneity, commodified into what he calls (in the title) a "Polynesian Hong Kong":

> it's a hootenanny
> and a hoedown
> if you're on da top
> and you pull da strings
> on all da puppet clowns.[49]

Such satirical tones reflect a global capitalist framework of "Asia-Pacific" with Asians on the Rim portrayed as not belonging to alternative frameworks (such as Oceania) or claims (such as those of indigeneity). Instead, this Pacific Rim version of Asia was aligned to tactics of domination, simulation, and interest for the transnational cultural class, a kind of "Disneyfication" writ large for global tourist consumption. as Fredric Jameson allegorizes this postmodern "ethnicity-effect."[50] In essays from "Imagining Asia-Pacific" (2000) to the chapter on Hauʻofaʻs carnivalesque Asia Pacific in *Be Always Converting* (2009), I resorted to antagonistic formulations such as *"Asia/Pacific* and *inter-Asia* do not just belong to the 'imagined community' of transnational capital and the astronaut class of [Asian] frequent flyers; they cannot just sublate Pacific into Asia."[51] So much more is taking place, and taking hold, in Asian and Pacific dynamics of cultural production and site-based work across decolonizing Oceania, and Asian sites and urban peoples, too, are "becoming oceanic," as is the broader claim here.[52]

In postcolonial Taiwan, to invoke a strong counterexample of an oceanic and greening island territory in the Northern Pacific, a school of archipelagic cultural studies work is arising that links Taiwan native studies to Native American transnational frameworks of outer-national and "trans-indigenous" belonging, on the one hand, and to a primordial and contemporary connection with oceanic frameworks that would unsettle territorial ties to the Chinese mainland and reframe this decentered island site as long connected to the Pacific Ocean, on the other. In an essay on these oceanic ties in Taiwan as refracted through works by the Tau poet Shyman Raporgan, from Orchid Island (long part of Austronesian culture and site of anti-nuclear protests in the 1980s), such as

Cold Sea, Deep Passion (1997) and *Black Wings* (1999), Hsinya Huang urges, "Through their own lived experience, as well as that of their island kin, Hauʻofa and Raporgan conceive of Oceania as a communal (sea) body, through which they can ultimately resist the imaginary political lines drawn by colonial powers. Their narratives turn hyper-modernized Pacific islanders (like themselves) back toward a perception of bodily identities as individual projects in intimate connection with Oceania." Huang notes that Raporgan, like Hauʻofa, "represents Oceanic peoples as custodians of the sea, who ʻreach out to similar people elsewhere in the common task of protecting the seas for the general welfare of all living things.'"[53] Huang invokes Raporgan's work on Oceania in northeastern Pacific spaces as an archipelagic region that reshapes the island nation of Taiwan into a trans-indigenous nation-space linked to Austronesian (if not Polynesian) modes of language, space, body, and culture: "What does the ʻworld atlas' mean? A chain of islands in Oceania. The islanders share common ideals, savoring a freedom on the sea. On their own sea and the sea of other neighboring islands, they are in quest of the unspoken and unspeakable passion toward the ocean or maybe in quest of the words passed down from their ancestors."[54]

In the ecological prefigurations of this transpacific nexus in *Earth House Hold*, Gary Snyder ends his poetic-didactic journey out of Cold War US formations and into alter-worlding constructions of place, self, and community in Asia and the Pacific. He links sites in Japan, India, Tonga, "Cold Mountain" China, and the Pacific Northwest by building up the Banyan Ashram on Suwa-No-Se Island in the Amami groups of islands that continue from Okinawa and the Ryukyus to Taiwan. This ashram, presided over by the prescient poet and transpacific dharma Buddhist wanderer Nanao Sakaki, cultivates ties to place, custom, and ocean through small-scale farming and fishing, "offering shochu to the gods of the volcano, the ocean, and the sky" and in oceanic bonding for nourishment, "For some fish you must become one with the sea and consider yourself a fish among fish."[55]

Meditating, farming, fishing, dancing, chanting, getting married to person and place, Snyder and his wife, Masa, and their mentor, Sakaki, push their transpacific journey toward an ontology of wider world belonging, situating Japan in an Oceanic framework: "It is possible at last for Masa and me to imagine a little of what the ancient—archaic—mind and life of Japan were. And to see what could be restored to the life today."[56] Snyder will work to bring this ecological and ethnopolitical stance I have been calling ecopoetics back across the Pacific in works of global and local eco-affects such as *Mountains and Rivers without End, Regarding Wave,* and *Turtle Island,* along with the volume of ecological-poetics essays *A Place in Space.*

Based in the Hawaiian islands, Albert Saijo's post-Beat, pidginized, "vandalized," Zen- and Emerson-haunted ecological rhapsody, OUT-SPEAKS, published with Bamboo Ridge Press, forges what he calls an alternative "cosmovision" of place, ocean, and planet from his residency on the edge of the Pacific near Hapu'u Forest in Volcano on the Big Island.[57] The kōlea, or golden plover, becomes his figure of an oceanic traveler living on scraps and edges, who, "A Kona," forges at once a line of flight and a mode of frugal inhabiting in (139–45). Identifying not as an ethnic or abjected Asian settler but as a "REBORN HUMAN" (197) of remade world ecology—the whole book is written in the declamatory capital letters of the manifesto ("I WANT TO OUTSPEAK" forces of capitalist domination like a "FIELD PREACHER" [17]) or trans-species rant, as in "Animal Rhapsode" (18)—Saijo (1926–2011) urges his credo of the small and caring life of hiking and sustenance living as beautiful. As he summarizes his quasi-Emersonian cosmopoetics and life dwelling close to the wilderness and his poetic and Buddhist quest (in the mode of Snyder and Lew Welch) for embodied beatitude in self and world, "EDGING AN ACTIVE VOLCANO—LIKE THEY SAY IF YER NOT LIVIN ON THE EDGE YER TAKIN UP TOO MUCH SPACE" (199). Saijo's poems, such as "O Muse," invoke and honor this "RADIONCARBONIC" and "BIOLUMINESCENT" oneness of body (bios) with the radio waves, carbonic presence, and light of world (13). In 2000, Elepaio Press and Hawai'i Dub Machine productions (in the Dharma Brothers Studio) released the compact disc *Jan Ken Po: Live in Honolulu,* which records the reading of poetry by these "transpacific dharma wanderers"—Gary Snyder, Albert Saijo, and Nanao Sakaki—whose work from the 1950s on has forged alternative visions of Asia and the Pacific tied to ecopoetic modes of planetary belonging, linking Buddha and animal with the human.

As discussed in chapter 1, book-length poems by Craig Santos Perez have gone on to enact an innovative, serial, and historically informed feat of repossessing Oceania and the Marianas as a mode of world belonging in which Guam/Guahan can never be named (or forgotten as) just another unincorporated territory of the post-1898 American Pacific.[58] It also can no longer be taken as just a militarized island with (as Robert Duncan saw it) "planes [forever] roaring out from Guam over Asia," turning the Americanized Pacific into "a sea of toiling men," "a bloated thing" of war, dispossession, and exploitation.[59] Such project-driven poems would proliferate counter-naming and trace routes and roots on Guahan, resulting in a counter-geography of archipelagic belonging to Oceania and the Marianas as more than an act "to prove the ocean / was once a flag."[60]

Santos Perez "convenes," as his poet mentor Aaron Shurin at the University of San Francisco and City Lights Press frames it, "an oceanic poetics": the poems, like their rooted and routed people, must begin again in saltwater and subsurface groundings and waterings, tracing "one saltwater" across different parts of the Pacific. "What the map cuts up," as Michel de Certeau puts this quest, "the story cuts across," as the poet works in a diaspora of open-field or circumoceanic poetics to tell the broken tale, in shards, remainders, space-time constellations of place, family, and hand-me-down stories.[61] It's Oceania as reconvened to put the water-land nexus back into pre- and postcolonial focus via a resurrected spatiality of four languages.

As Santos Perez writes, miming intertextual poetic and scholarly borrowings from Charles Olson as well as from Hau'ofa while composing his oceanic "field composition" poems, "Hau'ofa draws our attention to an oceania, preoceania, and transoceania surrounding islands, below the waves, and in the sky—a deeper geography and mythology."[62] Santos Perez does not just say this "New Oceania"; he enacts this oceanic-archipelagic region in a *performative worlding*. He quotes from the amazing Maori poet Robert Sullivan's waka "Ocean Birth": "Every song to remind us / we are skin of the ocean."[63] And he mimes the fluid documentary style of the Depression-era eco-documentary poet Muriel Rukeyser, from "The Outer Banks": "All is open. / Open water. Open I," making fixities break down and fuse, link across imposed divides of subjected verb, making "open" into world-making and I-breaking action.[64]

Becoming oceanic is not just subject matter or site but a poetic practice and ecological ethos, enacting transfigurative alterations of language, world, and consciousness; a mode of *reworlding* the island-ocean-city world. The Asia of these complicit poems by Santos Perez reveals an exploitative one, as well, across his middle-class diaspora, as well-off pregnant South Koreans arrive to give birth to children who become guaranteed US citizens as promoted by "birth tour agencies."[65] Postwar tourists begin to pour in from the Pacific Rim, particularly Japan, with its ties of war and colonial settlement in the Pacific Islands as in its own colonial Orient of an Asia for Asians: "1967: 109 passengers on pan am flight 801 from haneda, japan arrive; 'japanese rediscover guam,'" as "ginen sourcings" grimly puts the timeline, by 1973 a quarter-million tourists come to Guam, 70 percent of them Japanese.[66]

The very numbered sections of the poem all have Japanese numbers embedded in them, along with English and Chamorro and Spanish, *ichi*. The crafty rebranding of Guam as "world class tourist destination" and hotels "all with ocean views" continues as a function of what Teresia Teiawa troped as the "militourist" mode of space-production of the Pacific for Asian and Euro-American fulfillment.[67] Even as the grandmother's Catholic rosary ties the Pacific together in grassroots beatitude and oceanic crossings, Santos Perez writes, "When I say rosary [in Chamorro] I think I can hear her voice / even here in California."[68] The "guma" collection (2014) continues to extend the historical, linguistic, and formal range of Santos Perez's ongoing serial poem, *from Unincorporated Territory*, pushing this "fatal impact" narrative into quasi-"epic" scope as a portrayal of Guam in its indigenous, colonial, modern, and contemporary complexity. The contradictory island site is portrayed as at once hybrid-assimilated and decolonizing. Ethnography, autobiography, history, geography, literary criticism, journalism, ecology, the fascist writings of Ezra Pound, the spectral figure of Juan Malo, dietary regimes of junk food such as Spam, as well as poems by contemporary Chamorro poets, collide and get woven into this seriocomic intertextual mix of oceanic convening.

No Asia or Pacific region-making framework can remain innocent of these uneven power dynamics, historical elisions, bordered exclusions, and internal discrepancies of its own aporias of place making. Postmodern Oceania (full of Sullivan's Maori wakas and Santos Perez's Chamorro sakman, as well as huge, 350-meter-long containerized diesel ships from Matson and Evergreen, with such ships losing some 10,100 containers each year at sea) offers no eco-cure or postcolonial kava pill for the Cold War hangovers of war; post-9/11 remilitarization; racial tension; or the dynamics of neoliberal globalization reshaping space, time, self, and world.[69] We cannot forget racialized war, colonialism, and neocolonial economic discrepancies in the magical new waters of the Pacific becoming Oceania.

Still, affiliation to Oceania can become not just a matter of heritage or blood, but a mode of practice helping to forge "a trope of commitment, vision, and will"—hence, the remaking of Asia and the Pacific.[70] As Gayatri Spivak reminds us, reading literature from such discrepant spaces and peripheral cultures of smallness and rural distance, "The old postcolonial model—very much India plus the Sartrian Fanon—will not serve now as the master model for transnational to global cultural studies on the way to planetarity."[71] For we are dealing with planetary heterogeneity, she argues, in sites such as post–Soviet Russia and (what she rightly calls) "other Asias," across a decolonizing Oceania, emerging on a scale of intervention and responsibility more than just reflecting the same old, tired global-postcolonial hybridity fix as a repetitive postmodern mode.

Toward an Ecopoetics of Transpacific Solidarity via "Oceania"

Across eight postwar decades and transpacific contexts, Gary Snyder has forged a coherent ecopoetics, from *Earth House Hold* to the present, as gathered in his presciently titled collection *A Place in Space: Ethics, Aesthetics, and Watersheds*. Snyder is a poet activist coming down from the Beat era who has long advocated the regenerative power of wilderness, what he calls "the practice of the wild," and enacted deep ties of the Pacific Rim to the powers of emplaced consciousness and reinhabitory energies in the wilderness contado. In Snyder's geo-poetic reframing of the coastal Pacific into what, in his essay "Coming into the Watershed" (1992), which has become crucial to the field of American ecological criticism, he calls, "The San Francisco / valley rivers / Shasta headwaters bio-city region" are all interconnected (by slashes here) and lead to mode of gratitude and planetary care for "Turtle Island" as ethical attitude.[72] Snyder renames this biocommunity from his home in the Kitkitdizze Sierras bioregion the "Shasta Nation."[73] There, regenerative energies of the wild and primordial planetary belonging can lead Euro-Americans, Asian Americans, African Americans, and North Beach dharma bums on the reworlding path to "become 'born-again' natives of Turtle Island."[74] Snyder's ecumene is ecologically interconnected, planetary, and re-nativized counter-conversion to place.

In "North Beach," his essay on San Francisco urban place in *The Old Ways: Six Essays* (1977), Snyder enacts an uncanny bio-poetics of the region as contado and as counterhistory and counterculture. North Beach is portrayed as a "non-Anglo" multicultural habitat where the Costanoan native peoples had lived for more than five thousand years that later became a place of Alta Californian dairy farms before waves of Irish, Italian, Sicilian, Portuguese, Chinese (Kuang-tung and Hakka), and "even Basque sheepherders down from Nevada" settled in.[75] Beneath the Transamerica Pyramid, Snyder conjures up "a tiny watershed divide at the corner of Green and Columbus," where "northward a creek flowed" toward Fisherman's Wharf, all covered by oblivious landfill now. Evoking remnants of the Pacific bio-region and the occluded history of settlement, Snyder aims at "hatching something else in America; pray it cracks the shell in time."[76] That some-thing else is a vision of the Pacific bioregion that sees place connected to watersheds of bioregional belonging, place-tied values that come down to ecopoetics from Native America and global cultures of Asia and the Pacific.

In "Indigenous Articulations," James Clifford reaches into "articulation" theories of Stuart Hall and Antonio Gramsci on the coalitional forging of counter-hegemony to offer a multi-edged model of Pacific region-making he calls "subaltern region-making."[77] Pacific indigenous peoples can cre-atively compose," in this process, "a region cobbled together, articulated [with global forces], from the inside out, based on everyday practices that link islands with each other and with mainland diasporas."[78] As in his early work on the "Melanesian world," Clifford turns back to Jean-Marie Tjibaou in New Caledonia and the Loyalty Islands, "where a composite 'Kanak' identity was emerging in political struggle."[79] Such an Oceania-based vi-sion of place, land, and identity as "inter-dependent" would "also embrace the Pacific sea of islands—a wider world of cultural exchange and alli-ances that were always crucial for Tjibaou's thinking about independence as inter-dependence," as Clifford summarizes Tjibaou's gesture toward the island land and sea ("Mais, c'est la maison") as world home in Oceania.

We can thicken the meanings and tactics of "Oceania" via a well-situated anthology from Auckland and Honolulu titled *Whetu Moana: Contemporary Polynesian Poems in English* (2003), in which ten Hawaiian poets figure prominently, and many of their poems are concerned not just with links to the people of "the 'āina" but also to sustenance from, connec-tions to, and wayfaring across "Oceania" (including ecological-oriented poems based in Hawaiian waters, such as "Spear Fisher" and "Da Last Squid," by Joe Balaz, and poems of Native Hawaiian ecological recovery by Brandy Nalani McDougall).[80] Crucially, in 1976, and in waves of Pacific-crossing voyages afterward, the Hawaiian voyaging project Hoku'leia had begun to remap and reconnect the Polynesian triangle across Oceania and helped to create this interconnected system of ocean crossing and star guidance via native knowledge, techniques, and community building cutting across nations and colonizing prejudices, as in Robert Sullivan's waka assemblages in *Star Waka* and counter-geographies and indigenous ocean-sailing tactics of Santos Perez in *from Unincorporated Territory*.

Thinking with and beyond Hau'ofa's crucial new vision of the Pacific, Oceania can generate the framework for the forging of ecological solidarity and the site of alternative modes of Asia and Pacific, or Pacific and Asia linkage and knowledge formation. The imagination can prove helpful as a mode of transforming social and regional practices and help prod the making, shaping, and gathering of what I have been calling ecopoetics. Literature can help us to see such links and affects among ocean, self, and planet. Like geographical cultural studies, poetics can help overcome what Lawrence Buell has called "the foreshortened or inertial aspect of [the] environmental unconscious" so we can develop better modes of re-inhabitation and a "watershed consciousness" of an Oceania ethos aware of our ties to rivers, tidal shores, and the global commons of the ocean.[81]

Taking Care of the Body across Oceania: Sailing Olap's Micronesian Canoe

Using Clifford's evocation of New Caledonia as connecting place to ocean world; Snyder's oceanic ashram in Japan with Nanao Sakaki in *Earth House Hold* and *Jan Ken Po*; Saijo's ocean-facing Big Island in *OUT-SPEAKS*; Sullivan's eclectic waka assemblages; and Huang and her colleagues' renativizing turn of Taiwan into a counter-mainland site aligned to Oceania, I have aimed to overcome the taken-for-granted view of an Asia Pacific imaginary with Asian cultures and sites cast as transnational capital forces of globalization set against the interior Pacific figured as raw resource, fantasy site, vacancy, or subaltern or labor. Aiming toward a multiple-edged vision of ecological solidarity in the region, "We [culture workers, critical theorists, teachers] can seek the antagonistic synergy of Asia/Pacific forces, flows, linkages, and networks."[82] With wry wit and capacious-hearted humor, Hau'ofa often implied as much in his first-person-plural evocations, as when he left that catholic "we" of oceanic solidarity open, underspecified in the summary title of his selected works *We Are the Ocean*, and capable of coalition building inside and across the Pacific and the world. We are the ocean, indeed, in some fundamental ecological sense of body and world.[83]

As Sylvia Earle wrote in *Time* magazine in 1996 (in an article Hau'ofa was fond of citing), "Every breath we take is possible because of the life-filled life-giving sea; oxygen is generated there, carbon dioxide absorbed. . . . Most of Earth's living space [its ecumene], the biosphere, is ocean—about 97%. And not so coincidentally 97% of Earth's water is Ocean."[84] We know from extreme weather effects such as El Niño and polar melting, the sea shapes weather and climate patterns, and its moisture stabilizes and replenishes the fresh water of rivers, lakes, and streams. We are the Ocean in our bodies, as well, each living person composed of some 60–70 percent water. As Milton Murayama, the Japanese Hawaiian author from Maui, advised in *All I Asking for Is My Body* (1975), "Take care da body" and you may just take care of the place, soul, and other creatures on Earth as well in this world-making process.[85]

This chapter has pushed "Oceania" toward a Pacific-affiliated ecopoetics of translocal solidarity, place, and bioregional worlding being built up across the Pacific Ocean that writers such as Sakaki, Snyder, Westlake, Richard Hamasaki (alias red flea), and Saijo have long stood for in their "transpacific dharma wanderings."[86] The ocean that we are part of will know, and the planet will feel, our consequences for generations to come. One last "ecopoetic" image I would invoke to linger as a cautionary note: in a section titled "Flows," in *Asia/Pacific as Space of Cultural Production*, Dirlik and I included a translation by Theophil Saret Reuney of an ocean-based aboriginal work from Truk (in the Federated States of Micronesia), "The Pulling of Olap's Canoe."[87] The work itself, and footnotes with gaps of untranslatability, has names for birds, whales, plants, waves, rocks, navigation customs, islands, species of Oceania, as in the lines "The whale whose names are Urasa and Pwourasa / They guard those pompano fish which belong to wasofo [a name for the new canoe, and by extension the new navigator]."[88]

Reuney's works are now being cited by linguists and biologists to compile Chuukese names for plants and animals, some of them by now extinct.[89] The cultural studies scholar Joachim Peter is citing Reuney's works to forge an oceanic-based vision of horizon, world, and place.[90] Let us hope that these names and these creatures can survive our own planetary plundering across the Anthropocene. But the world of Olap's ocean is endangered and full of ecological pathos, as when Reuney's footnote 48 to the line, "You delve deeply into the fish of mataw anu," records that for the fish name *mataw anu*, the "meaning is ambiguous, especially since the type of fish is unknown."[91]

FOUR

Migrant Blockages, Global Flows

*Worlding San Francisco in a Global-Local
and Transoceanic Frame*

Glamorizations by the tourist industry notwithstanding, travel and
migration for the vast majority of people have been and continue
to be unhappy if not catastrophic occurrences brought about by
unhappy if not catastrophic events: the Middle Passage, the Spanish Expulsion, the Irish Potato Famine, conscripted military service,
indentured labor systems, pursuit of asylum.

· NATHANIEL MACKEY, *Splay Anthem*

Migrant Blockages, Global Flows across Europe
and the Americas

As Thomas Nail predicted in his far-ranging sociological study *The Figure of the Migrant*, and the African American vanguard poet Nathaniel Mackey prefigured in the poetic tropes, gaps, puns, and stanzas of
mobility, flight, disjunction, and fugitivity in his "blutopic" serial poem
Splay Anthem, "The twenty-first century will be the century of the migrant."[1] The figure of a billion migrants moving precariously, in these
hyper-globalized days, across borders, across oceans, across national and
regional frontiers, often torn from homelands and cast into refugee sites

or temporary encampments, gives the lie to trickle-down economics and liberal care. Such human figures are fleeing economic crisis, the eruption of ethnic intolerance, rural disintegration, war, and environmental disaster. Worlding aspirations might well encounter stupefying disorientation, if not demoralization and blockage. Disoriented by the 24/7 dynamism of deworlding globalization, as well as what has come to be called *compassion fatigue*, hardly captures the demands on transnational empathy or the anxiety of urban or national fear before such forces at once social, environmental, and fateful.

Still, in some rooted sense, the figure of the migrant arises as phobic synecdoche for the late capitalist system in its reach toward allowing transnational porousness and financial expansion (as well as labor blockages), sublime in all this dangerous instability and risk and hard to totalize or to cut a way into via literary, critical, or filmic imagination across lands, cities, or oceans.[2] The end of immigration myths, abjected and blocked by forces of global reaction by the forty-fifth US president, Donald Trump, or related forces in Poland or Germany and the virulent forms of white nationalism spreading across Europe and the United States, signifies the end of social imagination, democratizing energies, and the telos of globalization in progressive senses as a world-transforming force. Those remnant figures of humanity fleeing the planetary disaster of a frozen Earth, on the claustrophobic train around the locked-in earth in Bong Joon-ho's dystopic film *Snowpiercer* (2013), might well serve as prefigurative warning of the environmental fate, technoscience blundering, and the "climate refugees" to come as the planet heats and freezes, melts down and clogs up. The sixty thousand South Koreans waiting to reunite with their relatives in the communist North begin to suggest the long-term historical impact of wars, hot and cold, on such migrants and refugees and their families, values, divisions, and life patterns, from Syria and Palestine to Guatemala, the Marshall Islands, and South Africa.

In the wake of ongoing wars from the Middle East to Africa that shatter modern nation-state borders and policies, the new Europe faces what has been called "the largest migrant and refugee crisis since the end of World War II."[3] In the past five decades alone, under the shift to a state-capitalist model of production and financialization geared to global and transpacific distribution, the People's Republic of China has experienced the biggest rural-to-urban migration in human history and shaken up the peripheral spaces of Taiwan and Hong Kong with the threat of one-party nationalist absorption.[4] In a contiguous region, military expansion, political movements, and economic displacements all factor into the shaping of current South Asian migration patterns under US power and the rise of China across and beyond the South China seas as its so-called geo-historical oceanic domain.[5] This *specter* of fugitive complexity, vague threat, and liminal sublimity continues to haunt the triumph of late capitalist globalization across the Pacific Ocean and into other world oceans, from the Mediterranean to the Atlantic and the Persian Gulf: the specter of the migrant troubles borders and closures of nation-state, race, religion, and region.

Migrants as a global category must now include what the far-ranging eco-critic Rob Nixon calls the rise of "developmental refugees" and so many "uninhabitants" forced from their places of rural, local, or national belonging by disruptive mega-projects (such as big dams and nuclear power plants) and impacts of "slow violence" that result not just in the displacement of human dwellers, but in the uprooting of the land itself and ocean as resources of nourishment, health, and sustainability. As Nixon argues about what we have been calling *deworlding* forces, such environmental displacements commonly result in what he calls "a loss that leaves communities stranded in a place stripped of the very characteristic that made it habitable."[6] Blockages at borders of national space and the modernizing thresholds of a development-driven future of mobility haunt the present amplified global flow and its dangerously reconstituting "ethnoscape" of speed, flow, and mix. Indeed, migrant flow implies the migration of meaning into precarious embodiment and confronts the capitalist sublime where the categories of identity representation and civic advocacy break down and the taken-for-granted language of the nation-state all but fails or falls away from such problems into evasion and reactionary enclosure, as in the phobic US slogan "Build That Wall," which targets Mexico and scapegoats Latin America as a caravan of illegals, rapists, and psychopaths. We are living inside what has been called, from the pro-market as well as anti-Irish, anti-German, and anti-Catholic years of Ralph Waldo Emerson to the anti-market Leninist provoker Slavoj Žižek, "the new opacity" of transnational border bashing, state policing, racial phobias, and authoritarian bullying and the scattered rise of nationalist re-nativization.[7]

When as a turn away from the hopefulness of Beats, hippies, and communards, with their Mexican blankets, oatmeal cookies, Third World sandals, and grass, Bob Dylan lamented in his *John Wesley Harding* album song from the 1970s, "I pity the poor immigrant, / who wishes he would have stayed home," this might seem a comforting sentiment in these days of austerity cuts, raw racialized hatred, and precarious displacement—as if there still is a home, job, stable environment, or land of peace and community to return to (in places such as Syria, Palestine, or Iraq). Few citizens are immune to the economic push-pull dynamics of downsizing, extraction, capital flight, homeless blight, or expulsion. Those "Chimes of Freedom" Dylan had boasted about in a prophetic song by that title from 1964, with all the certainty of his American Jeremaic redemptive imagination bespeaking social change and economic uplift into carnal liberty, do not seem to apply anymore inside the global system. Dylan's chimes of freedom and calls to movements of peace and liberty are seldom, or no longer, "flashing for the refugees on the unarmed road of flight / And for each an' every underdog soldier in the night."[8]

We have come a long way from those upward-bound Euro-American immigrant networks and melting-pot dynamics that once welcomed the "migratory alien," or "trans-nationals," as Randolph Bourne called them in a prescient *Atlantic Monthly* article from 1916 wherein he hailed the rise of a new "Trans-National America," with its "spectacle of the immigrant refusing to be melted" into the dominant Anglo-Saxon settler model but, instead, often cultivated the "literatures and cultural traditions of their homelands."[9] What Bourne hailed as the "spiritual welding" of a cosmopolitan America out of this social process as one open to immigrant difference and transnational influx is nowadays becoming blocked at the migrant borders to Mexico, Canada, and transpacific Asia. We now face policies of a terror-haunted world of blockage and tariff, reactionary abjection and mean-spirited isolation. The figurative melting pot of the Americas, as in the European Union (EU) of unstable political and economic migrancy, risks fast becoming what we could call Fortress America, as signaled by Trump's phobic call for a Rio Grande River–tracing wall erected like a prison enclosure between the United States (in) and Mexico and El Salvador (out), as well as scapegoating, undertheorized retreat from transpacific free-trade agreements with China (for which he blames the "Wuhan flu," currency manipulation, and technological piracy) and countries of the Pacific Rim in a region such phobic discourse would rename the "Indo-Pacific."[10] The Biden regime at times echoes these bleak policies.

It is no longer the more transatlantic-centered case, as Karl Marx claimed in post–*Communist Manifesto* contexts, that "the United States [has] absorbed the surplus proletarian forces of Europe through immigration."[11] Any contemporary right to "global citizenship," as Žižek warned in a critique of Michael Hardt and Antonio Negri's *Empire*—with its grassroots "communist" embrace of mobile multitudes moving across national borders from below, the influx of cheap, mobile labor from the periphery of the global system—"would result in a populist revolt against immigration," which has already begun to happen in sites such as the EU-exiting Great Britain, France, Germany, Sweden, and the border states of the United States.[12] But this migrant figure, in the European social imagination and policy, has long been shadowed by phobic imagery and metaphors of "mass invasion," as Saskia Sassen warns in *Guests and Aliens*, her study of European migration patterns, causes, historical legacies, figura, and policies across two centuries.[13]

Given amplified conditions of global risk, technological mobility, and regimes of liquid financialization under the transnationalizing system, Zygmunt Bauman contends, "All people may now be wanderers."[14] The "debt-saturated and increasingly deregulated financialization" of markets that began in the 1980s, as David Harvey observes, has facilitated "geographical mobility" and the dispersal of labor and flights of capital across national and regional borders that have aggravated those "crises [that] are essential to the reproduction of capitalism."[15] Under such conditions of crisis, "illegal alien or '*sans-papiers*'" become what Harvey calls "a separate population vulnerable to unthinkable and unrestricted exploitation by capital."[16] It is not just labor that is moving across borders and oceans but capital itself under neoliberal conditions of transnational financialization and mobility. "Even when capital does not migrate," Harvey reminds us of this neoliberal embrace of markets as save-all *pharmakon*, hurting and healing peoples and places by capricious turns, "the very threat that it might do so often serves to keep labor quiescent in its demands."[17]

Any "homeland" nation becomes an opaque mesh of globalizing-localizing forces that disturb frameworks of nation-state modernity and national canons of self-representation and oceanic control. Political refugees, a precariat thrown from homeland sites, as well as economic migrants moving across borders legally or otherwise flow into opaque figurations of the global multitude. This is not just a reconfiguration of landed or settled polities. Migrancy is also a *world-oceanic* transaction, we should recall, a cross-water and transoceanic flow often forgotten in our Earth-centric figurations of migrant motions. This is the case in the flow of migrants from Syria to Greece and Turkey and across the Mediterranean Sea from North Africa into Italy and Germany. So-called sea slaves of oceanic labor spread across world seas and are exploited on "ghost ships" remote from or unseen by those prosperous cities that eat the cheap fish they help provide or feed it to their pets at home.[18] This forced labor in sites such as the South China Sea goes unseen or gets sublimated into the global process of forgetting the ocean in the latest seafood dinner in Hong Kong, Seoul, or San Francisco. The United Nations High Commissioner for Refugees (UNHCR) has put the number of refugees and migrants who arrived by sea in 2015 at more than one million people, with the majority (80 percent) landing in Greece, particularly on the legendary Lesbos Island, with others crossing the multi-harbored Mediterranean from North Africa into Italy. Warshan Shire, a trenchant British Somali poet, reminds privileged urban readers in London and New York City of these treacherous oceans: "You have to understand / that no one puts their children in a boat / unless the water is safer than land."[19]

In an earlier dynamic of *push-pull* cross-border flux, Sassen traced this to "seasonal migrations" and proletariat labor migrations across Europe since the turmoil of 1848 and, all the more so, to mass dislocations of civilian millions from the two world wars in the twentieth century; migrants are at times welcomed as *guests* (treated as potential insiders) or dealt with as *aliens* (marked as perpetual outsiders by phenotype, religion, language, or culture).[20] The formation of the European interstate system and the creation of independent nation-states, Sassen contends, has maintained an "intimate connection" with social categories of "the refugee, the displaced person, the asylum seeker" as manipulated by the state's right to determine citizenship, to affix social entitlements, and to enforce exclusions.[21]

The crucial paradox haunting our era of amplified globalization is that neoliberalizing countries (such as those in the EU, as well as the United States) "have had to negotiate the tensions between economic policies that lift border controls for the flow of capital and migration policies aimed at strong border controls for peoples."[22] Along these lines, the United States fails to acknowledge that, over the past five or more decades, this polity has been neutralizing borders to transnationalize its economy at the same time it has been closing borders via immigration policies of migrant blockage oblivious to human rights. Two different regimes exist, Sassen argues—"one for the circulation of capital and one for the circulation of immigrants"—putting this dilemma of contradictory migrancy at the core of contemporary nation-states and at times rendering their policies incoherent or unjust.[23] Indeed, this remains one of the fundamental contradictions of a system that is at once more liquidly *open* to capital and yet vigilantly *closed* to cross-border labor flows. Trump is just one sign of this white phobic closure and compassion fatigue, as if he never exploited migrant labor in his real estate in Queens or Florida or anywhere on this watery Earth riddled with golf courses and homeless camps.

Myriad tropes of *traveling* and figurations of cultural-political *displacement* have emerged, across the past decades, to express the ambivalent ingredients and historical discrepancies into the uneven making of some "imagined transnational community."[24] Migrants have become mobile figures embodying these far-flung crises of dispossession, austerity, displacement, and instability.[25] As specters of temporary "denizenship" (more than aspirant citizenship), they haunt the global circulatory system and reveal its unjust and latent polarities of aliens versus guests.[26] Free-trade figurations are haunted by the unfree migrant presence shorn of human rights and dignity, from Australia to Hong Kong and Poland. Xenophobia and racism trouble the dream of global mobility and liberal rights that mark utopic models affirming globalization from below, as in Hardt and Negri's porous *Empire*, with its call for rights to global citizenship that affirm the creative mobility of labor and the precariat such as at the US-Mexico border.[27]

"Diasporas" of economic migrants and "exiles" such as the political refugee leak into one another and suggest weakening ties to the myth of diasporic return and lure of the homeland, resulting in explanatory paradigms of globalization-*cum*-national regulation that seem at once belated and overextended.[28] James Clifford still offers the most comprehensive portrayal of such "diasporas" as a global reframing of migrancy: "For better or worse, diaspora discourse is being widely appropriated. It is loose in the world, for reasons having to do with decolonization, increased immigration, global communications, and transport—a whole range of phenomena that encourage multi-locale attachments, dwelling, and traveling within and across nations."[29]

Such diaspora-centered figurations of the migrant often ignore prior claims to indigenous priority in the face of such amplified mobility—in sites such as Hawai'i, Papua New Guinea, and Taiwan, for instance. It does not do to conflate immigrants and refugees into "one broad category of [diasporic] people coming from the East and the South, basically driven by economic need," Sassen contends, when the causes are geopolitically motivated and diverse and the consequences can be long term and disruptive.[30] Etymologically, *refugee* derives from the Latin verb *fugere*, meaning "to flee back to" or (more so nowadays) "to flee from" a political or environmental crisis, whereas *migrant* derives from the Latin root for *migration*, meaning a flexible "change of place or dwelling site," often in response to seasonal shifts or in quest for jobs or resources of social well-being elsewhere.

Conflations of refugee, migrant, and nomad all too often ignore or sublimate into misrecognition of social or class factors of discrepant mobility, location, security, environment, and prosperity to make claims along generalizing social lines, such as "We are all postmodern migrants now." Such claims about transoceanic and cross-border movements recall vapid turn-of-the-century Deleuzian figurations such as "We are all postmodern nomads," or the postcolonial platitude "We are all diasporic" and in-between subjects of hybridity. We are not talking here about Bohemian vagabonds, nomads of deterritorializing quest energy, postmodern textualists, or culture-seeking cruise ship tourists all collapsed into the subaltern figure of *the migrant*. Even Nail, a wary sociological empiricist, seems to push toward such sweeping claims when he suggests, in his conclusion, "In this sense, the figure of the migrant is not a 'type of person' or fixed identity but a mobile social position or spectrum that [most] people move into and out of under certain social conditions of mobility."[31] Most migrations are limited in scale and duration; moreover, they are historically caused and patterned, and they do end, as Sassen has shown in discrepant European contexts.[32]

Still, this *migrant-refugee* amalgamation has become a contemporary figure for the disruptive sense of social and material displacement and risk, so many human beings cast on the road, across borders, and across seas to further depths of insecurity, statelessness, bare life, exclusion, and secondary status.[33] They face conditions of "permanent structural inequality" that give the lie to political claims of equality, universality, or liberty that, say, the Surrealists in City of Light Paris or the Beats in earlier eras of American postwar affluence had presumed.[34] Border crossings aggravate claims to human settlement and universal state rights. Still, as Michel Foucault warned in an interview given to a Japanese publication in 1979, wherein he predicted that the plight of displaced boat people set across oceans from Vietnam and Cambodia presaged the great migrations of the twenty-first century we now face, "One should not remain indifferent to historical and political analyses of the refugee problems, but what needs doing urgently is to save the people who are in danger."[35] It is hard to disagree with such a humanitarian plea, although we do need to grasp what forces the migrant embodies in conditions of aggravated mobility and world risk.

To grasp the *deworlding* and *reworlding* dynamics of such discrepant social flows of migrants across the Pacific, the fictional works and essayistic interventions of the "transpacific" novelist-scholar and literary force Viet Nguyen (and other emergent transpacific diasporic writers his collections have highlighted) can be invoked and studied to overcome longstanding US historical amnesia and create social empathy across Asia and the Pacific for border-crossing figures of Asian refugee status and American immigration into states such as Texas, Arizona, Washington, and California that American wars and racially phobic security tactics have long helped to produce.[36]

Anachronistic Figures and Migrant Frames

Paul Gilroy warns in *After Empire: Melancholia, or Convivial Culture*, "The postcolonial migrant needs to be recognized as an anachronistic figure bound to the lost imperial past. We need to conjure up a [more convivial] future in which black and brown Europeans stop being seen as migrants. Migration [theory] becomes doubly unhelpful when it alone supplies an explanation for the conflicts and opportunities of this transnational moment in the life of Europe's polities, economies, and cultural ensembles."[37] For Gilroy, the long duration of the migrant figure incoming from global peripheries to the metropolis is haunted by imperial and colonial legacies of racism and urban or state exclusion, such that claims to multicultural assimilation and neoliberal diversity can hardly account for the rise of contemporary conflicts, phobias about otherness, and exclusions in European states and cities as we move toward the making of what he utopically evokes as a "convivial postcolonial urban world" of coexistence and "mutable, itinerant culture" as the future.[38] Gilroy goes on to warn, in the related study of postcolonial globality *Postcolonial Melancholia*, "We must be careful about returning to what we can call a 'migrancy problematic,'" whereby the culture and the insufficient modernity of the migrant remains the core explanation of European problems with border-crossing mobility.[39]

"Cultures of migration," with dreams of upward mobility and transformation at home and abroad, are affected by distant shifts and closures.[40] Blocked and militarized borders are less the exception than the rule.[41] The refugee figures as, or often stands in via unconscious synecdoche for, a kind of living social death in a regime of perpetual motion coming to be taken as a neoliberal constant of global capitalism—a socially kinetic movement from dwelling or worlding to life to a globalized disruption of place and nationhood without security, status, citizenship, or benefit. The examples could be multiplied from Hong Kong and Tokyo to Berlin and London. Rural-to-urban migration patterns (both within and across borders of nation-states, from Africa to Germany or, in the rural-urban US register, from Chiapas to Watsonville and Fresno) are huge, omnipresent, and disruptive of policies and patterns that have long governed immigration regulation and control. "Algorithmic justice" no longer proves adequate to mapping this influx from formerly rural country to megacity. As Annette M. Kim, director of the Spatial Analysis Lab at the University of Southern California, in Los Angeles, summarizes the South-North migration patterns in sites such as Ho Chi Minh City, Saigon, and Manila: "For the first time in human history, the majority of human beings live in cities. This shift has occurred because of urbanization in the Global South, which has primarily been driven by the influx of rural migrants. Despite coming to cities with fewer resources, these new urbanites have helped fuel the urban economy by providing a large body of low-cost labor. At the same time, anti-immigrant sentiment has caused growing contestation of their right to be in the city, pushing newcomers to eke out livelihoods in precarious situations." Top-down urban planning cannot match, fit, or locate shifting settlements, such that "some of the poorest residents are often living in precarious places without public sewage, drainage, solid-waste, and other infrastructure systems."[42]

The migrant does not so much explain as trouble and haunt enduring American, emergent Asian, or time-honored European models of pluralist assimilation and liberal citizenship. The migrant as a temporal being lives in-between stable polarities of *dwelling* and *belonging* (recognizable in the Dylan songs I invoked earlier). As Nail argues of the migrant's *in-between* state of deferral, motion, and unrest, "The 'emigrant' is the name given to the migrant as the former member or citizen, and the 'immigrant' as the would-be member or citizen. In both cases, static place and membership [as state citizen] are theorized first; the migrant is the one who lacks both."[43] Refugees do not want to be forever called "refugees," as Hannah Arendt has argued in World War II contexts: "We ourselves call each other 'newcomers' or 'immigrants'"—that is, empathetic figures of uprooted humanity with the right to claim rights.[44]

The migrant figure, living between these polarities of being sent (*displaced*) and being or becoming received (*settled*), is coming to be recognized as a dominant social figure in the unstable transnational present of world-altering transformations. Postcolonial literature, for example, has moved from an emphasis on adjudicating claims of *exile* (in writers such as George Lamming and Theresa Cha) to more unstable or vaguer claims of *migrancy* (in writers such as Michelle Cliff and Jamaica Kincaid), with Salman Rushdie and Bharati Mukerjee embracing the ambivalent and deconstructive term *(im)migrant*.[45] Walter Benjamin has been said to be such a dispossessed European *migrant*, blown by historical forces across Europe, living precariously on bohemian edges as a writer and communist, later cast into cross-border flight and suicide as a Jewish refugee from Nazi machinations in Germany and France.[46] Arundhati Roy has called the US security-state whistleblower Edward Snowden a transnational "refugee" because "he cannot return to the place he thinks of as his country (although he can continue to live where he is most comfortable—inside the internet)," as if he dwells in Russia but lives on the worldwide web, just as the Australian computer expert Julian Assange lives not in Russia or the United States but in exile on Wikipedia and was incarcerated in London since April 2019, but he has now been released.[47]

Refugee flows into Europe and across Asia have been aggravated by US wars in Indochina and Central America, not to mention by the post-9/11 wars in the Persian Gulf. Even the on-the-road automobile and highway nexus so beloved of the mobile US Beats cannot be disconnected from what Nixon calls the "slow violence" of environmental damage and migrant displacement in distant yet interconnected sites and peoples. As Aldo Leopold had reminded overconsuming Americans, critiquing a dynamic of environmental profiteering all the more aggravated by recent wars in the Persian Gulf, "When I go birding in my Ford, I am devastating an oil field, and re-electing an imperialist to get me rubber."[48] These not-so-distant homelands of migrant flux and global ties haunt the American ever-urbanizing present. The Dominican American novelist Junot Díaz was castigated as "anti-Dominican" and stripped of his 2009 Order of Merit award for advocating for the right of "undocumented immigrants" (mainly border-crossing Haitians) to live in the Dominican Republic, the Caribbean country from which his family had migrated to live in the United States in 1974.[49]

In an era of trickle-down Reaganomics and multicultural and gender experimentation in sites from Brownsville, Texas, to Santa Cruz and Watsonville, California, Gloria Anzaldúa's *Borderlands / La Frontera* poetically (in a vanguard sense of worlding transfiguration) prefigured this condition of living across and between violent borders (between North and South, from Texas and California to Mexico) as a "herida abierta" or open wound "where the Third World grates against the first and bleeds" in the Reagan era that has endured into the regime of President Joseph Biden.[50]

Toward Positive Refigurations of Migrancy: Worlding San Francisco

Situated on the edge of the coastal Pacific and blessed with human, as well as environmental, resources from the Sierra river-systems to the grandeur of the San Francisco Bay and Mount Tamalpais, San Francisco needs to be seen as a vanguard global city collocating forces and resources (financial and creative) on the Pacific Rim. It is a nexus that consolidates the finances, resources, neighborhoods, creative-destructive dynamics, and techno-cultural creativity of the Silicon and Fresno valleys, as well as of Napa and Marin counties, to the north and east. Taunting federal immigration blockages of the Trump administration, San Francisco (like Los Angeles, which initiated de-federalizing urban-sanctuary policies in 1979) prides itself on being a vanguard US "sanctuary city," resistant to federal immigration laws and policies since the City and County of Refuge ordinance passed in 1989 and accommodating to cross-border flows of peoples, cuisines, labor, technological resources, and cultures from the Americas and from across the transpacific (Asia and the interior Pacific), such that, for example, more Tongans now live in the San Francisco contado of San Mateo than in Tonga itself.

The sanctuary space of solidarity for the transpacific migrant inside San Francisco has not always been easy or flexible, then or now, from the days of white nativist anti-Chinese attitudes and frontier laws; through the days of US wars, hot and cold, in Asia, which included the internment of Japanese American citizens in World War II; to the continued instability of borderlands / la Frontera, where some would like to erect a "Great Rio Grande Wall" to keep so-called migrants from Mexico and Central America out.[51] The shared if differently situated experience of US white nationalist racism was one of the unifying factors that provoked the coalitional making of Asian America as such, bringing twenty-three distinct ethnic groups in the United States together into an uneasy political fusion. Even after the Immigration and Nationality Act of 1965 dismantled long-standing racial quotas, Erika Lee argues in *The Making of Asian America*, Asians in America were considered "perpetual foreigners at worst, probationary

Americans at best."[52] Affirmatively stated in the present tense, Lee pushes us toward a conclusion; as readings of exploratory literary works such as *Tripmaster Monkey* and *I Hotel* show by the end of this chapter, "These transnational immigrants are helping us all become global Americans."[53]

I consider Maxine Hong Kingston's novel *Tripmaster Monkey*, along with Allen Ginsberg's *Howl*, Jack Kerouac's *Dharma Bums*, Karen Tei Yamashita's *I Hotel*, and Barbara Jane Reyes's *Poeta en San Francisco*, Punk-based memoirs by Michelle Tea, and mapping works by Rebecca Solnit, to be masterpieces in portraying global-local complexities of San Francisco on the edge of corporatized America. *Tripmaster Monkey* opens with the image of Wittman Ah Sing—a fifth-generation Chinese American in San Francisco's Chinatown, whose immigrant ancestors go back to the days of the nineteenth-century Gold Rush and transcontinental railroad building—contemplating suicide from the Golden Gate Bridge, as if that golden gateway between Asia and continental America and Europe had led not to "gold mountain" but to ruination, misery, failure, and death: "The last city. [He would jump] Feet first. Coit Tower giving you the finger all the way down. Wittman would face the sea. And the setting sun."[54] The Pacific is figured at the outset as a space not of crossings but of limits and ends, as Jack Kerouac portrayed it in "October in the Railroad Earth": "End of the land sadness end of the world gladness all you San Franciscos will have to fall eventually and burn again."[55] Rejecting suicide and walking through the North Beach of the Beats and literary romanticism, Wittman spots "a Chinese dude from China, hands clasped behind, bow-legged, loose-seated, out on a stroll—that walk they do in kung fu movies when they are full of contentment on a sunny day" (4–5).

Wittman may be more melancholy—or suffering postcolonial melancholy, as Gilroy would diagnose it—but he is no belated American migrant. He is a fifth-generation California citizen who belongs to San Francisco as an artist: "Wittman stopped dead in his tracks, and shot the dude a direct stink-eye. The F.O.B. stepped aside" (5). The FOB (Fresh Off the Boat) is like an SOB to Wittman here: inferior, raw, smelly, disgusting, unassimilated, an eyesore. "F.O.B. fashions—highwaters or puddlecuffs. Can't get it right. Uncool. Uncool" (5). He is "uncool" and square in North Beach, the bohemian home of Beat cool and City Lights, where Wittman felt he belonged with the other would-be Kerouacs and Bob Kaufmans. Wittman will end up crossing the Pacific over to China and back to San Francisco Chinatown, as a performance artist, to stage his drama of the monkey king become Tripmaster monkey, railing against America as against China and Chinese Americans to forge his own freewheeling satirical and poetic art as a kind of refigured activism and dramatized intervention into those forces that would drive him to abjection, self-hatred, blockage, melancholia, and death. Beat ethics of ecological simplicity remained crucial to his quest for a mode of poetic living and ethical world belonging in the manner of Lew Welch and Gary Snyder: "Wittman [Ah Sing] was not much attached to stuff, trying to live by The Red Monk's advice that fifteen things are too many. Be open-handed; be free. Let the bookstores and libraries take care of the books. Read them and give them back or away. To be free in America: rid yourself of impedimenta."[56]

Karen Tei Yamashita's experimental novel of far-ranging scope *I Hotel* is also set in San Francisco Chinatown and on the multicultural and bohemian fringes of North Beach, that "non-Anglo" space of cosmopolitical mixture, as Gary Snyder boasted of this Beat milieu in 1975, as "habitat . . . the world moves in and through" and hatches "something *else* in America."[57] Yamashita's tenfold novel is not just named for the International Hotel (or "Eye-Hotel," as it opens); she retropes it as the "I-Migrant Hotel" for the year 1974. She tracks, with archival dizziness and multigeneric capaciousness, the 1968–77 struggle to make Asian migrants belong to place, city, and country even as place, city, and country belong to them in ways that go beyond, or more deeply into, the place than do the Beats on the road to world bohemian sites. The book opens in atmospheres of death in 1968, a year of murders; the year that "our black daddy Martin with a dream and our little white father Bobby will take bullets to their brains," leaving poetry-loving Paul Wallace Lin (named after Wallace Stevens) and other sons of Chinatown "monkey orphans."[58]

Like *Tripmaster Monkey*, *I Hotel* opens with death—not the near-death experience of Wittman, but the death of Paul's father, a painter from "the Monkey Block" (12) of modern artistry, from an early heart attack on the streets of San Francisco Chinatown, amid television news reports of mounting deaths in the Vietnam war. This home city can seem a foreign, war-torn, site too, even to a Lowell High School or San Francisco State University (SFSU) student: "It's not that Paul has never left Chinatown, but it's always foreign out there. Chinatown his citadel" (8–9). Revolution by students and workers is in the air, as Saigon and San Francisco get overlaid via scenic juxtaposition, death and war abroad, and death and peace at home. Still, when you do die in San Francisco and are buried outside in Colma, you find peace and a sense of belonging at last: "Too late for immigration to deport you" (11).

Revolution and Maoism are in the air of this 1968 San Francisco Chinatown, from Berkeley to Paris and Beijing and back. The racial coalitions that go on strike at SFSU and the University of California (UC), Berkeley and lead, at the very least, to large-scale institutional and racial changes and the implementation of ethnic studies become in the reactionary semantic and cynical retelling of the leading Japanese American S. I. Hayakawa a way to divide and conquer the racial groupings: "Establishment of the department came with some fanfare and a budget just substantial enough to create a sensation of power and competition, creating political fissures between black, brown, and yellow, throwing into contest what had once been idealized as a rainbow of colored solidarity" (21).

This "Chinatown Verité," in the layers of immigration and contestation Yamashita portrays, makes it belong to Asian Americans: "We looked like the enemy [after Pearl Harbor and the war in Vietnam], but that's not the same as being the enemy. . . . We weren't tourists. We lived here" (58–59). As the chapter on "I-Migrant Hotel" portrays through the voice of the Filipino worker and I-Hotel resident Felix Allos, the rural "grass roots" and urban "brick roots" (433) are transnational "routes" that extend back to the Ilocano labor struggle in the Philippines and to the plantations of Hawai'i (430), as well as into Delano struggles in Central California (423–24), where transpacific Asian laborers and activists joined with Chicanos to fight the mega-agricultural corporations, as they would to fight the transnational corporations coming in from Hong Kong or Thailand to take over the International Hotel and evict them and abolish their history of im/migrant belonging to San Francisco.

Yamashita narrates sections of the "Eye-Hotel" in a "we" collective voice that shifts from representing Japanese American ("We gave our adopted towns names like Li'l Yokohama and Nihonmachi" [58]) to invoking Chinese American history ("Looking closely we saw the wood [on the walls at Angel Island] was carved away, inscribed in Chinese characters" [62]) and back again in the same chapter section, in an uneasy, unstable first-person-plural coalition rife with antagonisms and protests that go back to earlier struggles between the Kuomintang and Mao Zedong and Japan and forward to the struggles over the Tiao Yu Tai islands, as well as the Cold War antagonism between communism and capitalism being fought out at home and abroad—not to mention the tension to institute ethnic studies at the two Bay Area universities, with differing class formations and student bodies, as well as "liminal" ties to homelands.[59] The novel ends in this collective "we" voice reflecting "our transmission from the International Hotel" (579), in a defeat that implies victories of Asian American activism from such ruins: "We saw the morning sun rise over our I-Hotel in a sea of rubbish: broken glass, splintered wood, horse shit, lost pieces of clothing and shoes, abandoned scarves, face masks and helmets, torn posters, paper flyers and shredded banners, blood and vomit" (587).

The figure of the transpacific migrant haunts the ruins of *I Hotel* with the specter of future victories and rebirths, like that rising sun over the ocean and that blood, sweat, and tears of cultural-political activism. As the Taiwanese scholar of Asian American and inter-Asian literature Chih-ming Wang argues, "Through political activism and literary activities, Chinese students since the 1970s, like their predecessors sixty years ago, have brought Asia and America together to pursue the ideal of national independence and to explore the possibilities of critical internationalism, albeit while holding on to different and alternating national imaginaries and identifications as Chinese, Taiwanese, and (Asian) American."[60] Erika Lee phrases this victory from such struggles on the US front in these social terms: "Out of the 1960s a new Asian America was formed. . . . [It saw] the widespread involvement of Asian Americans in a number of campaigns for civil rights, women's liberation, lesbian, gay, bisexual, and transgender (LGBT) rights, and an end to the war in Vietnam. Out of this political participation came the emergence of a distinctive Asian American movement that helped to define diverse peoples as Asian Americans and call them to action."[61] The San Francisco–based but globally reworlding novel *I Hotel* in effect helps us to see how this making of an Asian America was aligned, rooted and routed, transversally entangled and interarticulated, with so many other migrant, labor, and human rights struggles, national and international, weaving a transpacific and trans-Americas frame around Chinatown/Manilatown/Japantown/world San Francisco that makes us remember this spectral past.

Works of empathetic imagination such as *Tripmaster Monkey* and *I Hotel* reveal an inventive urban worlding and global reworlding of locality and urban dwelling. They figure forth and articulate global-local social dialectics that need to be read in such transpacific migrant and coalitional contexts. They show how moving peoples differentiate and aggregate, split apart, then coalesce across differing generations; ethnic and political situations; and dominant and emergent configurations of social power, laboring creativity, struggle, and activism in radiantly prefigurative and thick-descriptive ethnographic, social, and historical ways. These narrative texts are *worlding* narratives that unpack the sublime "opacity" of the *migrant figure* and make us actively see and hear—bringing into conditions of audibility, if not implied advocacy—these global flows and racialized ties and wounds that bind and liberate across the Pacific. Both of these richly local urban works of multicultural world literature can help us theorize/figure forth/unblock the social complexity and politics of "the migrant" as a global figure situated in and beyond any American-centric or Asian American frame of San Francisco in the world. Such works help to give the specter of the migrant the depths and textures "positive refiguration," a social transfiguration from exclusion to prolonged inclusion that is much needed in urban transoceanic sites from Africa, France, Italy, Greece, and Germany to Taiwan, Korea, Hong Kong, Canada, and California.[62]

As Donna Haraway writes about the troubled, crisis-ridden Anthropocene, we live in a time when, "right now, the earth is full of refugees, human and not, without refuge."[63] Urban food hubs, such as the wholesale produce market in San Francisco's Bayview Point, along the port harbor, are threatened by rising tides and extreme weather events (prefigured by hurricanes Sandy and Katrina) that render urban living for the poorest residents, from Boston and New York to San Francisco and Long Beach, less secure and even more precarious. Creative works of worlding scope such as *Tripmaster Monkey* and *I Hotel* (as well as *The Last Black Man in San Francisco* [2019], a place-haunted movie of African American liquidation directed by Joe Talbot, and the San Francisco street-based poetry of Tongo Eisen-Martin in *Heaven Is All Goodbyes* [2017]) are radiant research probes into the sublimity-from-below of the mobile migrant, unblocking characters such as Wittman and Felix Allos and their colleagues for social captivity and mute abjection into world humanities recognition, social empathy, refuge, acceptance, and global transformation. The migrant blockages cannot prevent the reworlding dynamics of these global transoceanic figures.

III

TRANSPACIFIC
CONJUGATIONS

Unmaking and Remaking Worlds

FIVE

Under a Golden Gate "Mushroom Cloud"

Urban Space, Ecological Consciousness,
and the Pedagogy of Blue Conversion

Our oceans bear the brunt of our plastics epidemic—up to 12.7 [million] tons of plastic end up in them every year. . . . Municipal bag, cup and straw bans like those in Morocco, Iceland, Vancouver and some US cities [such as San Francisco] are a great start, but also not enough. And while clean-up efforts [such as recycling] are helpful in addressing litter problems, they can't begin to touch the problems created by microplastics—the tiny participles of plastic that now permeate our waterways and broader environment.

· ANNIE LEONARD, "Our Plastic Pollution Crisis Is Too Big for Recycling to Fix"

The war of the worlds hangs here, right now, in the balance it is a war for this world, to keep it a vale of soul-making.

· DIANE DI PRIMA, "Rant"

Starting from the Coastal Pacific

This chapter grows out of a large undergraduate Literature Department course on the geomaterial formation and literary-poetic heritages of San Francisco that I have taught every three years since 2003 at the University of California, Santa Cruz (UCSC), where I moved back across the Pacific to teach cultural studies, postcolonial literature, and creative writing in 2001. I returned to the coastal state of California (where I had been educated at the UC Berkeley from 1967 to 1976) after twenty-four years teaching American poetics and Asia Pacific literatures in the local/indigenous-challenged (as well as reenergized and dialectical) English Department at the University of Hawai'i, Mānoa, and at various universities across the Pacific Rim, including Korea University and Korea National University of the Arts in Seoul; National Tsinghua University in the tech-savvy hub city of Hsinchu; the oceanic-facing National Sun Yat-sen University in the port city of Kaohsiung, Taiwan; and the postcolonial and perilously decolonizing English Department at the University of Hong Kong, where I taught "Asia Pacific Cultural Poetics in the Era of Global/Local," filled with graduate students across disciplines, from music to American studies and comparative literature, and from countries as diverse as the People's Republic of China (PRC), Hong Kong, the United States, Australia, Taiwan, France, and England.

The coastal, oceanic, and surfing history city of Santa Cruz, situated on the Northern Pacific coast seventy-five miles south of John Steinbeck's "The City" (San Francisco), to which world-literary roads in California would lead for metropolitan culture and social transformation, has long been part of the greater Bay Area contado. Affiliating worlding poesis practices with what is coming to be called climate urbanism across the Silicon Valley region between San Jose and the mountains and coast to the north and west, *contado* is a term I will invoke and thicken to define and tie together (via the UC Berkeley urban geographers Gray Brechin and Richard Walker, among others) San Francisco and Santa Cruz in critical, environmental, worlding, and planetary ways.[1] In the larger dynamic of Left Coast–affiliated values of coastal Northern California, I agree with Diane di Prima that any "cosmological" world making in these precarious times of the Anthropocene is also a matter of soul making, given the deworlding hold of what Allen Ginsberg blasted in *Howl* as the death god of war and the ever profiteering god of profit, Moloch.

The aim has been to provoke a situated turn, or eco-conversion, toward a cultural-political way of thinking about worlding such Pacific Rim cities as complex environmental formations connected to common resources and shared, if imbalanced, issues of watersheds, oceans, agricultural areas, place, race, and social justice that cut across usual "urban-wilderness" or "country-city" divides that are coming undone in sites along the Pacific Rim as inside and across Oceania. We need to think both globally and locally, as Wimal Dissanayake and I have urged since the mid-1990s and Annie Leonard more recently warns about urban plastic-dependence patterns and their huge impact on world oceans in formations such as the Great Pacific Garbage Patch conglomerating offshore to comprehend global cities such as San Francisco, Honolulu, Taipei, and Tokyo.[2] As Leonard argues, "Recycling alone will never stem the flow of plastics into our oceans; we have to get to the source of the problem and slow down the production of all this plastic waste."[3]

During the Gold Rush boomtown days of San Francisco's financial and material growth after 1849, the coastal and forest-rich town of Santa Cruz provided abundant supplies of timber and limestone (at resource-rich quarry sites such as Cowell Ranch, which was later donated to form the basis for Cowell College at UCSC's opening in 1965) to help build up what Brechin elaborates as the financial wealth, resource extractions via mining, grand parks and hotels, and the European-mimicking architectural grandeur of "imperial San Francisco" and its far-flung contado hinterlands.[4] Santa Cruz would serve as a summer beach resort and surf culture boardwalk, from the 1880s to the present. Nowadays, ironically, Santa Cruz provides critical theory and the rise of cultural studies of "left-green" affiliation at the Center for Cultural Studies (a US version of the Birmingham School of Stuart Hall)—for example, in works such as *The Worlding Project*, published in 2007 by New Pacific Press, in Santa Cruz at the theory-rich Literary Guillotine Bookstore—along with more techno-cultural and micro-electronic contributions to the wealth and grandeur of Silicon Valley. Santa Cruz, as biographers and *Rolling Stone* magazine are fond of telling us, is the coastal bohemian haven to which Apple's founder, Steve Jobs, used to drive over Highway 17 to search flea markets and garage sales for bootlegged Bob Dylan cassettes. Jobs undertook this journey so he might continue to quote romantic and surrealist Dylan lyrics to his girlfriend, who dreamed only of his making enough secular money for them to live on in the outskirts of San Jose.[5]

Starting from Santa Cruz and San Francisco as global-local and transoceanic Pacific sites, I am again "worlding" these as discrepant US sites embodying and projecting distinctive transpacific cultural and historical meaning and growth. Worlding as a critical practice enacts an opening of space, time, and consciousness to other values and multiple modes of being, as we have discussed. Spatially, a worlded criticism seeks to disclose altered and emergent connections and articulations that cut across place, area, city, and regional forms, as I touched on in chapter 2. The left-leaning undergraduates in the humanities at UCSC do feel part of San Francisco and its "Beat" subterranean traditions and thus are glad to go to that Pacific Rim city on weekends to do site-specific research on its cosmopolitan influx and borderlands outreach at sites such as City Lights Bookstore, North Beach, the Tenderloin, the Castro, Union Square, and Haight-Ashbury.[6] Santa Cruz also informs part of a complex Bay Area formation that scholar-poets such as Joshua Clover, Jasper Bernes, and Juliana Spahr, coeditors of the radically anarchist Commune Editions of poetry and polemic in Oakland, have called "the Red Triangle" of the anticapitalist movements, protests, and riots emanating from UCSC, UC Berkeley, and UC Davis from 2009 to the present. The city and university of Berkeley were at once *antinuclear* in urban attitude and yet crucially involved, as university and US city, in the scientific-industrial production of the atomic bomb at the Lawrence Rad Lab and as part of the Manhattan Project. As Spahr and David Buuck write in their Occupy-affiliated *An Army of Lovers* (2013), what does it mean "to be poets in this time, this time of wars and economic inequality and environmental collapse, and in this particular urban space [reflecting on the border between Berkeley and Oakland], that put up signs claiming to be a 'Nuclear Free Zone' despite being the place that was largely responsible for the development of the nuclear bomb?"[7]

As the surrealist Afro-Jewish Beat poet Bob Kaufman had warned, from North Beach sites of cultural-political estrangement, in his mock "Abomunist" (absurdist communist) newscast writing and as collaged in his "Abomunist Manifesto" from the Cold War era of US / Soviet Union world polarity and threat of a global nuclear holocaust (still with us now): "End of news. . . . Remember your national emergency signal, when you see one small mushroom cloud and three large ones, it is not a drill, turn the TV off and get under it. . . . Foregoing sponsored by your friendly [Beat] Abomunist. . . . Tune in next world."[8] An "Abomunist" is an absurd surrealist communist—that is to say, his claims are hyperbole or rants not taken seriously as an oppositional force to the US war machine. Kaufman's manifesto (like Ginsberg's *Howl* and Richard Brautigan's *Trout Fishing in America*) was produced during a time when the US military and atomic commission was testing more than twenty-three nuclear bombs, which were detonated by the United States between 1946 and 1958 at seven test sites on the Bikini Atoll reef, on the sea, in the air, and underwater. We destroyed these Pacific dwelling places with long-term radioactivity and created what have been termed Pacific Islander "nuclear refugees" across the Marshall Islands to this day in that so-called Californian-void Pacific.[9] This is part of the American Pacific to which San Francisco, Berkeley, and Santa Cruz belong, not just the long-standing surf culture that ties these Northern Pacific sites to the surfing princes and diasporic *aloha* mores of Hawai'i, with which my research has been concerned as a pedagogy and ethos of situated intervention and the "becoming oceanic" poetics.[10]

Writing on the Left Side of the World: Abomunist San Francisco

The globally influential Beat poet Allen Ginsberg summed up hostility to nuclearizing military-industrial US culture when he cursed, in the meandering psychodrama of his midcentury poem "America":

> America, when will we end the human war?
> Go fuck yourself with your atom bomb.
> I don't feel good don't bother me.[11]

Ginsberg was suffering from what Spahr and Buuck, in the latter-day nucle-arized horizon of *Army of Lovers*, would call "being ill with late capitalism," as these post-Beat poet-activists try to chant, exorcize, meditate, curse, protest, raid, attack, or blow it up.[12] San Francisco enables and provokes such poetic hyperbole and techno-artistic edges of invention; venture capitalism; and a do-it-yourself (DIY) playground of adventure, risk, damage, and death at the end time and Pacific coastal cliffs of the continental sublime.

"[San Francisco has become] this far-out city on the left-side of the world": this is how Lawrence Ferlinghetti, the entrepreneurial founder of City Lights Bookstore, summarized San Francisco's impact on the world as a cultural-activist production site of counterculture, union-aligned, antinuclear, and queer radicalism in his inaugural address as the first poet laureate of San Francisco in 1998. In his jazz-inflected Beat poem "West Coast Sounds—1956," Bob Kaufman shows in wry imagery that the cultural miming of Black spaces and multicultural modes has been displacing San Francisco's ethnic enclaves via the growth of "San Fran hipster land," absorbing the hinterlands of art. This process has been aggravated by the ongoing absorption, or "death," of San Francisco into the techno-mores and modes of Silicon Valley as upscale real estate and gentrified couture. Kaufman, like some closed-down sardine factory in Monterey

Bay, splits for the global South and Mexico, if only to preserve his underemployed life and poetic vision of "cool beatitude":

San Franers, falling down.
Canneries closing.
Sardines splitting
For Mexico.
Me too.[13]

To continue with this worldly approach to literary and geopolitical San Francisco, I invoke such literary and cultural-political passages to give you a feel for the buildup of San Francisco as place of *creative-destructive* dynamics and urban-rural alteration. This is what film director Alfred Hitchcock (as well as the US Marxian poet George Oppen) would later bring forth, as San Francisco's disorienting "vertigo" effect on the environment. This effect was early portrayed by Bayard Taylor in *Eldorado*: "Now scarce a day passed [in the autumn of 1849] but some cluster of sails, bound outward through the Golden Gate, took their way to all the corners of the Pacific. Like the magic seed of the Indian juggler, which grew, blossomed, and bore fruit before the eyes of the spectators, San Francisco seemed to have accomplished in a day the growth of half a century."[14] Or consider this no less hopeful transpacific configuration from the later Beat era: "They [original San Franciscans] had their faults, but they were not influenced by Cotton Mather."[15] Here, Kenneth Rexroth is implying that San Francisco's polytheistic, bohemian, and anarchistic culture was not connected to that American social formation of witch-hunting New England Puritanism but had a transpacific edge.

In the wake of all that Wild West, bohemian, beat, hippie, queer, anarchist, and cosmopolitan experimentation in San Francisco from the Haight-Ashbury days to the multicultural present, it is still the case that, as the band Jefferson Airplane embodied in its "white rabbit" quest for utopian states of oceanic consciousness and a community in which love and peace would rule: "Somebody once said, if you want to go crazy, go to San Francisco. Nobody will notice."[16] The sad fact these days is that, given the high cost of living in San Francisco as it becomes a Google-bus metropolitan hub of Silicon Valley down to San Jose, Mountain View, Palo Alto, Los Altos, and Cupertino and connected to knowledge-power sites such as UC Berkeley and Stanford University, millennials are flocking to Sacramento, Oakland, and Watsonville, where their computer-*cum*-coffeehouse lifestyle is at least more affordable in the larger Bay Area.

The counterculture fascination of Silicon Valley's techno-creative forces with figures of left-leaning visionary redemption is perhaps best captured in historical anecdotes such as this one on the creative origins of Apple in Los Gatos and Cupertino: "[Apple's founders Steve] Jobs and Steve Wozniak initially bonded over their mutual obsession with Bob Dylan. 'The two of us would go tramping through San Jose and Berkeley and ask about Dylan bootlegs and collect them,' said Wozniak. 'We'd buy brochures of Dylan lyrics and stay up late interpreting them. Dylan's words struck chords of creative thinking.'" The Dylan bootlegs they acquired were mainly on reel-to-reel tapes. "'I had more than a hundred hours, including every concert on the '65 and '66 tour,' said Jobs. 'Instead of big speakers I bought a pair of awesome headphones and would just lie in my bed and listen to that stuff for hours.'"[17] In 1982, at twenty-seven, Jobs even dated Joan Baez, who still had that Dylanesque aura as the leftist protest muse of Carmel and inspiration for "Visions of Johanna" and "Farewell Angelina." Jobs would meet and have long meandering conversations with Dylan near Palo Alto in 2004.[18] To put this transfiguration of worlding poetics provocatively, the head and light shows of the Fillmore and Avalon ballrooms and the conjunction of Beat poetics at City Lights Press in Ginsberg, Ferlinghetti, and Kaufman, with the music and world literature (poetry) of Dylan, Jefferson Airplane, and the Grateful Dead would become internalized in the Apple laptop and iPhone and lead to world-transformative forces, mores, and forms.

Worlding San Francisco's Environmental Contado

San Francisco's vast urban periphery ("hinterlands" of the countryside) has long provided the material resources (water, timber, stone, agriculture, shipping, and so on), as well as the huge labor needs and creative inputs, to build up the wealth and splendor of the "imperial city" à la some Rome or Constantinople of the Pacific coast. This contado goes far beyond the smallish forty-seven square miles, forty-three hills, and 800,000 population of physical San Francisco city proper and connects urban well-being to the watersheds and back-to-the-land movements of the High Sierras to the north and Silicon Valley, San Jose, and Big Sur to the south, if not to the mineral and oil resources of Alaska and the current transnational assembly lines of the Pacific Rim and Mexico for resources and survival as an economy and as a life-sustaining bioregion.

To supplant existing San Francisco Bay watersheds and reservoirs, fresh lake water from the Hetch Hetchy Reservoir has been tunneled and piped 160 miles west since 1934 and now provides 85 percent of the water used by San Francisco residents. Brechin draws his environmental framework of "urban power" from postimperial Italy: "Italians, with their long experience with city-states [such as Rome], have understood this relationship [between city and countryside "hinterlands"], though more in economic than ecological terms. For them, the civilized world was a duality made of the city and its *contado*—that is, the territory that the city could militarily dominate and thus draw upon. This space provided the city with its food, resources, labor, conscripts, and much of its taxes, while its people (the *contadini*) received a marketplace and a degree of protection."[19]

Assuming that the power of empire making and global urbanization veils itself in sublimating mythologies and architectural monumentality, like the gold-plated San Francisco City Hall or the myriad urban monuments to the pioneers and Native Sons of frontier mining, Brechin's *Imperial San Francisco* goes beyond the spatial obviousness of urban borders and the quasi-leftist clichés of urban mythology and Bear Flag Republic and its lone pioneer-worshiping history. Instead, Brechin reveals the huge "displacement" effect and impact of this environmental contado and the resources of water and land needed, whereby the costs and damage of urban growth and urban wealth are displaced to the urban hinterlands and rural peripheries and down to later generations, as urban splendor pushes its ecological costs "downwind, downhill, and downstream" spatially, as well as "down-time" historically, as with the mercury-tainted rivers from the mining days that are still endangering life.[20] Working in the cultural critique tradition and post-Marxist leftist wake of Raymond Williams's *The Country and the City* (1973), Brechin denaturalizes and defamiliarizes taken-for-granted urban borders and urban-rural divides and discloses rural, watershed, transoceanic, and agricultural connections crucial to the growth and well-being of San Francisco as a wealth-producing city.

In *The Country in the City: The Greening of the San Francisco Bay Area* (2007), a related study that overcomes sublimating containments of nature and ecology by pastoral myths or lone-yeoman georgic delusions of liberal American studies after Leo Marx and Thoreauvian pastoralists, Richard

Walker refuses the "environmentalist dichotomy" or false binary of a "rural pure countryside" posited against an "unnatural ever-degraded city." Rather, to reveal the social relations and patterns of labor and profit that have shaped both in geomaterial interaction, we need to see what Walker calls "the city in the country" (as in those urban wine tastes in Napa Valley), as well as the "country in the city" (as in all those organic foods from Watsonville and the Salinas Valley and those farmers markets at the San Francisco Ferry Building and filling the Metreon Theater).[21] There is not a square mile of "nature" in the Bay Area that has not been worked over by human labor and fought for by green activists, as in the "urbanized greensward" of Golden Gate Park or the protected vistas of the Muir Woods and Mount Tam so beloved of Gary Snyder, Rebecca Solnit, and Tom Killian as Zen space. The Bay Area's "remarkable amount of open space—green, blue, and golden" has been fought for by a century of environmental protection policies now integrated into the fabric of urban development.[22]

In transpacific Taiwan, with its long-standing Hsinchu–San Jose nexus, contemporary transformations of the Bay Area connect the finance and creativity of San Francisco to the expansion into a Silicon Valley contado: San Francisco is considered the world's fourth most important city-region, so called, in the global economy (after New York, London, Tokyo). This is based on its maintaining some "edge" in the techno-knowledge economy as a global hub for servicing business and innovation and serving as a magnet to attract creative and talented people. The creative-destructive effects of this are still being measured, as in Jeremaic work such as Solnit's *Hollow City*, which tracks the "hollowing out" of artistic, bohemian, and cultural forces by the influx of yuppie technocrats.[23]

Coming into Transpacific Watersheds and Commons

Brechin's grim assessment of imperializing and the defense web and bomb-making Silicon Valley decades after *Trout Fishing in America* echoes Richard Brautigan's lost pastoral sense of thoroughgoing commodification of American wilderness nature into real estate and urban resource extraction: "No area on the planet is free from the process of global urbanization. Wilderness has ceased to exist."[24] Aware of such urban-rural and "outback" survival and Zen mindfulness tactics and contexts on both sides of the Pacific Ocean, and the forging of what has been called his "Pacific Rim community" by Timothy Gray and others (including me), from the Pacific Northwest and the High Sierras to the zendos and ashrams of Japan and India, Gary Snyder has forged a coherent eco-poetics of green and blue tonalities. He has written and acted on this world-making and ocean-crossing poetics from *Rip Rap* days of hippie grunge to the more upmarket New Directions publication *Earth House Hold* to the present, as gathered comprehensively in his visionary collection of prose and poetry, *A Place in Space: Ethics, Aesthetics, and Watersheds*. I have used *A Place in Space* in various courses at UCSC, from Beat Literature and the World to undergraduate creative writing seminar assignments on walking and writing, as well as on describing the water sources that sustain one's writing life. The world-traveling environmentalist Snyder, a poet-activist from that Left Coast Beat era of Ginsberg, Kerouac, Kaufman, and di Prima, has long believed in the regenerative and sustaining power of the wilderness kept wild if sensibly managed with care, or what he calls "the practice of the wild," maintaining the deep ties of the Pacific Rim city (from San Francisco to Kyoto) to the powers of emplaced consciousness and reinhabitory energies in the wilderness of "rivers and mountains without end" (the trope and genre on which he borrows and builds from is taken from Chinese and Buddhist aesthetics and ethics).[25]

In his geo-poetic reframing of San Francisco in the essay "Coming into the Watershed" (1992), which has become crucial to the field-imaginary tactics of American ecological criticism from Lawrence Buell to Rebecca Solnit, as "The San Francisco / valley rivers/ Shasta headwaters bio-city region," Snyder has them all interconnected (by slashes), leading to an exclamation mark of gratitude.[26] It also leads to transspecies care for "all sentient beings" as an ethical attitude of Buddhist empathy and an ethos and poetics affiliated to Native American world making to which Snyder has adhered from his origins along the Pacific Northwest coast and in the mountains of the Pacific Cascades. Snyder elsewhere calls this bio-community, built up over decades of labor and meditation as practiced and written from his communal home in the Kitkitdizze Sierras bioregion above San Francisco, the "Shasta Nation," where the regenerative energies of the wild and the deepening sense of planetary belonging can lead Euro-Americans, Asian Americans, African Americans, and North Beach dharma bums on a reworlding path to "become 'born-again' natives of Turtle Island."[27] Over years and contexts of aging yet staying young, Snyder has cultivated an ecologically interconnected, planetary, and re-nativized counter-conversion to depths of transpacific place and planetary being, a cosmology of world belonging, as di Prima advocated in "Rant," as well as in her life's work, from *Revolutionary Letters* to *Loba*.

In Snyder's watershed-tracing essay "North Beach," about that all-but-sacred Beat urban place of inhabitation, artistic event, coffeehouse, and publication, he imaginatively enacts a fully uncanny bio-poetics of the region and a counter-history and counterculture to imperializing San Francisco.[28] North Beach is portrayed as a deeply and dearly "non-Anglo" multicultural habitat where the Costanoan native peoples lived for more than five thousand years along and fishing in the bay that later became a place of Alta Californian dairy farms of Hispanic ranchers before waves of Irish, Italian, Sicilian, Portuguese, Chinese (Snyder notes they are Kuang-tung and Hakka peoples), and "even Basque sheepherders down from Nevada" settled in and enriched the place and people of San Francisco with their waves of immigrant labor and culture.[29]

Beneath the corporate Transamerica Pyramid high-rise, as if in its spectral shadow, Snyder evokes the Montgomery Block of artists and left-ist activists who had lived there decades before the Beats arrived. No less important, he unearths "a tiny watershed divide at the corner of Green and Columbus" where "northward a creek flowed" toward Fisherman's Wharf that is all but covered by oblivious landfill and pavement now. Here abides another lingering sign of "the Pacific beneath the pavements" of the commercial city. By evoking these remnants of the bioregion and the occluded history of multi-immigrant settlement in this Silicon Valley hub, Snyder aims at "hatching something else in America; pray it cracks the shell in time."[30] That something else is a subterranean and on-the-oceans-road vision of San Francisco that would see it connected to living watersheds and the rain and flows of the oceanic atmosphere; to a sense of bioregional belonging; and to the influx and ongoing transformation of place-tied values that come down from Native America and global cultures to the too oblivious digitalized city of Google, Facebook, and Zoom (all with headquarters in Silicon Valley and with many of their workers living "high on the hog" in the costly real estate of greater San Francisco). "Worlding" can come to mean what David Trend has argued it dominantly means—that is, living as well as working in "virtual" places held together by advanced online technologies and mediated genres and sites that take us farther each day from those "face-to-face" encounters that the Beats and the people of Chinatown, the Castro, Haight-Ashbury, and the Fillmore variously cultivated in the streets and in their homes, diners, coffeehouses, studios, and workplaces, as well as in those swinging bars and clubs of the so-called Barbary Coast for which permissive San Francisco (as with queer culture mores of the Castro) was famous.[31]

San Francisco's ties to the Pacific Sierras and the Northwest's Cascades are shown in Jack Kerouac's theopoetic novel *The Dharma Bums* (1958), which in its plotline enacts a kind of Buddhist dialectics shuttling between the "samsara" of San Francisco Beat urban life and poetic ecstasy in North Beach and Chinatown and the "nirvana" to be found in the mountains. This post–*On the Road* novel recalls the years and mores that both Kerouac (the Ray Smith protagonist) and Snyder (portrayed as the idealized Han Shan seeker, Japhy Ryder) had worked as fire watchers in the Pacific Northwest, at sites such as those Kerouac called "Desolation Angel" in his memoir novel by that name.[32] As the Kerouac character, thinly disguised as Ray Smith, writes of his retreat to this mountain-minded wilderness: "I wanted to get me a full pack complete with everything necessary to sleep, shelter, eat, cook, in fact a regular kitchen and bedroom right on my back, and go off somewhere and find perfect solitude and look into the perfect emptiness of my mind and be completely neutral from any and all ideas. I intended to pray, too, as my only activity, pray for all living creatures; I saw it was the only decent activity left in the world."[33]

Kerouac's hero of this off-the-road quest for a transhuman planetary mode of living is Gary Snyder (so-called Japhy Ryder), who had opened up this Zen mindfulness practice linking urban life to the regenerative energies and more natural values of a sincere life tied to the wild Pacific Northwest outback:

> Japhy [alias Snyder] leaping up: "I've been reading Whitman [the preface to *Leaves of Grass* (1855)], know what he means when he says, *Cheer up slaves, and horrify foreign despots*, he means that's the [Beat] attitude for the Bard, the Zen Lunacy bard of old desert paths, see the whole thing is a world full of rucksack wanderers, Dharma Bums refusing to subscribe to the general demand that they consume production. . . . I see a vision of a great rucksack revolution thousands or even millions of young Americans wandering around with rucksacks, going up to the mountains to pray . . . and also by being kind and also by strange unexpected acts keep giving visions of eternal freedom to everybody and to all living creatures.[34]

Japhy was a word-play name by Kerouac borrowed from a vulgar name used for Snyder in some North Beach circles. To Ferlinghetti, as he noted in letters to Allen Ginsberg and others, Snyder was "Jappy." This Cold War nickname reflected both Ferlinghetti's horror at the US Navy's use of nuclear bombs in Hiroshima and Nagasaki, which he viewed as racial crimes, and Snyder's own days at UC Berkeley, where he studied ancient Chinese and Japanese languages and poetics and became involved with Japanese Zen practices and modes of living in the natural world. Snyder cultivated a poetic mode of dwelling in the Banyan ashram or a Japanese zendo, as I discussed earlier in the context of his postwar oceanic life on islands south of Japan and in Kyoto.

As discussed in the previous chapter, Maxine Hong Kingston crossed the streets of San Francisco Beat culture in *Tripmaster Monkey* with lines of flight to spaces across the Pacific Ocean, including Hawai'i in *The Fifth Book of Peace*, where a pacifist and antiwar culture of protest against American wars in Korea and Vietnam and a simpler mode of living away from consumer demands could be created. Such transpacific practices and lines of flight outside US Cold War military-industrial mores were kindred in spirit with those of Gary Snyder, Philip Whalen, and, especially (for Kingston), Lew Welch and the Zen Buddhist counterculture. As she writes of this Lew Welch figure and a Wittman character such as Frank Chin, as well as about her husband, the actor Earll Kingston, who was a fellow English major at UC Berkeley when they married, "[Moving to Hawai'i,] Wittman [Ah Sing] was not much attached to stuff, trying to live by The Red Monk's [Welch's] advice that fifteen things are too many. Be open-handed; be free. Let the bookstores and libraries take care of the books. Read them and give them back or away. To be free in America: rid yourself of impedimenta."[35]

Literature, framed in and as a process of world making, can help us to see links between the city and the larger contado of the rural and un-cultivated wilderness and the endangered planet. Like geospatial cultural studies, poetics can thus help us to overcome and reframe what Lawrence Buell has called "the foreshortened or inertial aspect of [the] environmental unconscious," so that we can develop modes of "reinhabitation" and cultivate a "watershed consciousness" aware of ties to rivers, outer-Pacific spaces, and the global commons of the ocean.[36] Writers such as Karen Tei Yamashita, Gary Snyder, Jack Kerouac, Bob Kaufman, and Maxine Hong Kingston can help us to see the occluded and intercon-nected worlds of urban-rural belonging, in a capacious sense cultivated as the transpacific sense of planetary "beatitude." In this worlding sense, works such as *Tripmaster Monkey* and *I Hotel* give an expanded vision of cultural-political activism. They also prod us toward a planetary ecological consciousness necessary to survival.

My own sense of worlding as a transpacific humanities practice was deepened, in comparative terms, as a participant in the Nature's Blues: Literature, Ecology, and Environmental Justice international conference for the Association for the Study of Literature and Environment (ASLE)-Taiwan in July 2009, organized in vanguard "blue oceanic," as well as "radical green organic," frameworks of environmental justice and reworlding by Shiuhhuah Serena Chou of the Department of Foreign Languages and Literatures, National Sun Yat-sen University, Kaohsiung. This visit and talk about my work enframing the literature of San Francisco in such ecopoetic contexts followed an earlier visit to the same ocean-facing university to team-teach (with the trans-indigenous Native American and Maori scholar Chadwick Allen) the summer graduate seminar Discourse of Pacific Culture, under the auspices of Professor Hsinya Huang. These "always converting" visits and stays in the Asia Pacific sites of Taiwan and in its southernmost port harbor and shipping hub, Kaohsiung, further entangled my long-wrought worlding poetics into this port city. "Nature's Blues," as Chou elaborated the framework, implied not just the *blue* of the Pacific Ocean and the ponds, rivers, and streams that feed into it, but also a *blues* of global-warming damage, plasticene filth, overheating, acidic wastewater, and so on that were irreparably turning this shared planetary ocean into some ugly Pacific garbage heap.

During the summer of 2009, the cherished contado of Love River canals, the agricultural hinterland of this booming port city, the surfing and tourist beaches south of the island, and the Sandimen mountains of the Paiwan people and Marian churches shaped like boats and wedding halls, all opened up this pre- and post-Han place and its multilayered people to the *walking/waking/worlding poetics* I elaborate in the next chapter as situated in Seoul and Kaohsiung. After living on the islands of Hawai'i for more than twenty years, and in coastal California for a decade more, Taiwan resonated as an archipelagic world island of diverse languages and beliefs; of Taoist fortune-telling temples; of fisherman- and sailor-cherished sea goddess and Fujianese shaman Mazu and her myriad temples in the harbor just below campus; of the hybrid cuisine and "weird English" signs beckoning consumers all over this globalizing city, including a coffee shop named Bagel Bagel (with its Chinese-language training books and disks popular with expatriates) that was nearly impossible to find, cutting-edge high-speed rail bullet train, and richly encul-tured urban maze of streets and shops. All become linked in this blue-green hub city into a nexus of the agricultural region south of Taichung, Hsinchu, and Taipei. The final chapters of this study suggest ties of this "Pacific beneath the pavements" city with the forging of a transpacific worlding poetics into which my teaching-writing vocation became more deeply tied, linked up with and indebted to Pacific Rim cities of "waking in Seoul," South Korea, and the "pirate island" of Taiwan, as well as the postcolonial Hong Kong and postnuclear Japan.

SIX

Hiroshima Sublime

Trauma, Japan, and the US Asia Pacific Imaginary

On September 16, 1985, when the Commerce Department announced that the United States had become a debtor nation . . . ,
the money power shifted from New York to Tokyo, and that was the
end of our [American] empire [in the Pacific].
· GORE VIDAL, "Requiem for the American Empire"

We must come to see ourselves, Japan, historically as a very organic
part of the Pacific. No one has written such a study yet, but I hope
that gradually a history of Japan as a Pacific culture will be written.
· KENZABURŌ ŌE, "The Myth of My Own Village"

Atomic force first entered history in August 1945 with earth-shattering,
unforeseen, and, in many respects, uncontrollable consequences we are
still coming to terms with on the transnational and post-Soviet globe at
a time when the nuclear threat remains undiminished from North Korea
to Russia, as well as from unforeseen accidents such as the one at the
Fukushima nuclear power plant in 2017.[1] It has taken more than fifty
years, since the detonation of atomic bombs in Hiroshima and Nagasaki

by US military forces, for this "traumatic kernel" to sink in and register within the geopolitical imaginary. Indeed, the sublime *unthinkability* of this atomic catastrophic event has much to do not only with its unprecedented magnitude and force—the material/spiritual grandeur that I theorize after the language and novels of Kenzaburō Ōe (1935–2023) as "the Hiroshima sublime"—but also with its traumatic nature within the imaginary of the national US subject. What some postmodern critics now theorize (after Jacques Derrida, among others) as "the nuclear sublime" that was activated at Hiroshima remains one of the unimaginable, transmaterial grounds of a global condition that, paradoxically, can and must be reimagined, represented, and invoked to prevent this transpacific trauma from happening in post–Cold War history.

As an ethical and aesthetic mandate for the new millennium, the Cold War repression of Hiroshima within the American political imaginary needs to be symbolically confronted and undone at national as well as global levels. As Americans and as Japanese citizens of the liberal global order, we must mutually move beyond the Cold War situation of historical repression that had obtained in 1965, when Ōe lamented, "To put the matter plainly and bluntly, people everywhere on this earth are trying to forget Hiroshima and the unspeakable tragedy perpetrated there."[2] However traumatic, Americans and their allies must *try to remember this Hiroshima sublime* as a trauma of geopolitical domination and racialized hegemony across the Pacific Ocean. By thinking through and reimagining the techno-euphoric grandeur of this Hiroshima sublime, as well as representing the ideological complicity of ordinary Americans in their own sublime (as if raptured by technological forces of sublimity as manifesting and installing Patriot missiles as belated signs of their global supremacy) and ordinary Japanese (as citizens of the Empire of the Sun fascinated by self-sublation into zeros of solar force, as I will discuss) in the production of this nuclear sublime, we can begin to mutually recognize that a "postnuclear" era offers new possibilities and symbolic ties between America and Japan as Pacific powers united across Oceania.[3]

This postnuclear era emerges out of World War II freighted with terror and wonder as a double possibility, urging the globe at once toward annihilation, yet also toward transactional and dialogical unity at the transnational border of national self-imagining. The phobic masochism of the sublime can no longer operate in a transnational world of global-local linkages, although the technological sublimity of the Persian Gulf War had suggested otherwise, with its "sublime Patriot" missiles and quasi-nuclear landscapes lingering in the world deserts from Iraq and Afghanistan to Nevada and North Korea.

With the breakdown of many binary oppositions from Orientalist discourse and all but underwriting the Cold War as an ideological standoff between two superpowers, the Soviet-Union-become-Russia and the United-States-become-undemocratic, we must continue to "work through" the nuclear sublime as a symptom of the modernist will to technocratic supremacy across the Pacific now threatened by the rise of a militarized Global China expanding its military bases southward into the ocean. Recalled as a collective labor of mourning, the Hiroshima sublime can be said to emerge positively, thus, as a possible fantasy of postmodern redemption in which world war, the hardware of the nation-state, and the will to industrial supremacy are, in effect, over. The weapons remain, but the transpacific world of Oceania has been transformed by trying to render them inoperative, even obsolete, as icons of geopolitical terror. Yet a Doomsday Clock keeps ticking louder than ever inside our Anthropocene climate.

As an ethical and aesthetic mandate for the 1990s, the Cold War repression of Hiroshima within the American political imaginary must be symbolically confronted and undone.[4] However traumatic, as has been argued in this chapter, postnuclear Americans and their allies must *try to remember this Hiroshima sublime* as a trauma of Pacific domination and Asian abjection. As Ōe wrote in *Hiroshima Notes* and substantiated with political-cultural activism across Japan and the region, "The Hiroshima within me does not come to an end with this publication."[5]

As the subject of such modern nation-state complicity, the Japanese kamikaze pilot in J. G. Ballard's *Empire of the Sun* (1984) worships the "sublime body" of Empire, but he refuses to die, to dematerialize, to sublate into Earth and sun. He even refuses the boy's symbolic redemption into hero worship; confronting this dead body of the kamikaze pilot, "[Jim's] hands and shoulders were trembling, electrified by the discharge that had passed through them, the same energy that powered the sun and the Nagasaki bomb whose explosion he had witnessed."[6] This young Japanese pilot, bravely willing to annihilate himself into the solar and militant enterprises of "the Empire of the Sun," is the British boy's "imaginary twin," whose body will survive the war, survive death as conquering spirit (363).

For Ballard, interned by the Japanese in Lunghua from 1942 to 1945, the atomic bomb comes deeply—and even like some weird historical redemption—out of the collective imagination of Europe and the unresolved territorial conflicts of Asia. Each newsreel and war movie shown serves as a symbolic rehearsal for warfare, and even with the war ended, this heroic imaginary begins over again, forever unsatisfied and displaced, seeking new historical territory on which to stage the same battles, the same obsessions and hysterical symptoms of "the (impossible-real) Thing."[7] The Hiroshima sublime abides as the impossibility and failure of Asia Pacific representation to reach after and convey this unconscious thing, this death wish that seems to emanate so deeply from the war project of modernity. The suffering induced by this sublime icon of American supremacy cannot be elided from Pacific memory or through romanticized figures of Oceania as perpetual unity.

As a focus for symbolic and physical deconstruction, the image of the nuclear bomb at Hiroshima threatens the postmodernist will to symbolization, the will to go on, the will to make meaning of individual life. Represented in postwar texts as more rupture than fact, "Hiroshima" can be read as an image of what James Hillman calls "the-death-of-God-God": an unprecedented act in which the latent nihilism of Western technocratic reason is made manifest as world death.[8] Under this sublime of nuclear technology, mountains sigh and disappear into Tinkertoys; flesh and birds melt; skyscrapers become props for Superman in his comic book mode as a pop Zarathustra incarnating peace on a grand scale—as in *Superman IV* (movie), for example, disposing of American and Russian nuclear weapons and waste by transporting them to virgin regions of outer space.

Weirded out by the threat of world death, machinations of the Cold War, deceptions in the name of political idealism, and the abyss of ecological accident, generations of Americans once hungered for self-transcendence, then turned to acid, acid rock, sexual fusion, and cultural confusion. Postwar generations became white hipsters on the road from Hiroshima to God knows where, as if some India of the mind. Meaghan Morris, seconding the historical frame of technology developed by Fredric Jameson, Ernst Mandel, and others that would link cultural formations of the postmodern to new electronic- and nuclear-powered apparatus that emerged in the 1940s, has argued that "the postmodern era could be said to begin in 1945, at Hiroshima and Nagasaki. It begins not simply on the ground, under the bomb, but in the relationship between that ground, those people, and the pilot who could only ever thereafter confront the 'reality' of the bomb, those deaths, through an image, a film, a story, a representation, a reconstruction, a vestige, a simulation of what had, or might have, happened on the ground."[9]

This impossible-to-see, transtextual event of nuclear power at Los Alamos signals the rupture of the postmodern, the traumatic disappearance of the modern-war real into the imaginary and symbolic geopolitics of war: "The postmodern begins with an experience in which it is impossible to 'see' unmediated empirical reality and survive; an experience which would-be survivors, potential victims, can only evoke and express with images, metaphors, fictions and rhetoric which they must try to convert into actions to ensure that we may never know that 'reality.'"[10] The nuclear bomb may disappear into semiotic infinitude, as Morris suggests, rendering nature effectively over as a symbolic source of psychological solace and moral stabilization. What Masuji Ibuse called the "crazy iris" of ecological deformity proliferates from this nuclear ash.[11] Nature has been disrupted by the nuclear sublime as we enter the accelerated terrains of the anthropogenic era.

Through the awestruck perspective of Jim in war-torn Shanghai, Ballard suggests that the world war and the nuclear bomb, like the heroic pilot of Empire, come out of the boy's technology-haunted imagination, fueled by war movies, comic books, *Reader's Digest*, *Life*, newsreels, and the merchant seaman Basie and the other unconscious adults around him:

> When Basie and the men had gone, vanishing among the ruined warehouses on the quay, Jim studied the magazines on the seat beside him. He was now sure that the Second World War had ended, but had World War III begun? Looking at the photographs of the D-day landings, the crossing of the Rhine and the capture of Berlin, he felt that they were part of a smaller war, a rehearsal for the real conflict that had begun here in the Far East with the dropping of the atom bombs. Jim remembered the light that lay over the land, the shadow of another sun. Here, at the mouths of the great rivers of Asia, would be fought the last war to decide the planet's future. (357)

The nuclear bomb at Nagasaki has become a fantasmatic spectacle of sun, death, and immortality for Jim; it at once ends one war but, reimagined, seems to foreshadow the brilliancy and location of the next world war.

Before and after the battles and internment around Shanghai, the will to war goes on recharging itself in the Pacific through its very symbolization, as the representation of war necessarily turns heroic, sacrificial, patriotic, ideal. Ballard is exact and uncanny here in recording the elation and terror of Jim, as his fascination with an all-too-innocent American supremacy emerges heroically, and falsely, out of the war's ruins:

> Surrounded by this vision of all the abundance of America falling from the air, Jim laughed happily to himself. He began his second—and almost more important—meal, devouring the six copies of the *Reader's Digest*. He turned the crisp, white pages of the magazine, so unlike the greasy copies he had read to death in Lunghua. They were filled with headlines and catch phrases from a world he had never known, and a host of unimaginable names—Patton, Eisenhower, Himmler, Belse, jeep, GI, AWOL, Utah Beach, von Runstedt, the Bulge and a thousand other details of the European war. Together they described an heroic adventure on another planet, filled with scenes of sacrifice and stoicism, of countless acts of bravery, a universe away from the war that Jim had known at the estuary of the Yangtze, that vast river barely large enough to draw all the dead of China through its mouth. (308–9)

So begins the rehearsal for the next Pacific war, emerging from the ground of heroic fantasy, this time instigated by the American political imaginary:

> Feasting on the magazines, Jim drowsed among the flies and vomit. Trying not to be outdone by the *Reader's Digest*, Jim remembered the white light of the atom bomb at Nagasaki, whose flash he had seen reflected across the China Sea. Its pale halo still lay over the silent fields but seemed barely equal to D-day and Bastogne. Unlike the war in China, everyone in Europe clearly knew which side he was on, a problem Jim had never really solved. Despite all the new names that it had spawned, was the war recharging itself here by the great rivers of eastern Asia, to be fought forever in that far more ambiguous language that Jim had begun to learn? (309)

According to the uncanny images Ballard invokes in his narrative, "the Empire of the Sun" had been defeated by the sun itself become a nuclear bomb. "When Jim looks unconvinced [that the war had ended], the Eurasian explained: 'Kid, they dropped atom bombs. The Americans threw a piece of the sun at Nagasaki and Hiroshima, killed a million people. One great big flash" (295). Jim can only reiterate an affirmation of solar witnessing that comes as much out of his own war-fascinated imagination as out of Pacific history: "I saw it" (295). Ballard's complex of solar-centric imagery resonates with US ideology of its moral and technocratic superiority here. "It is an atomic bomb. It is a harnessing of the basic power of the universe. The force from which the sun draws its basic power has been loosed against those who brought war to the Far East," President Harry Truman had urged, with millennial awe, in a radio address explaining the Hiroshima bombing to Americans in 1945.[12]

The *nuclear sublime* erupts out of the mundane and vulgar contexts of modernity, as if eruption from the unconsciousness of British, Japanese, and American alike, bringing an end to the territorial battles that had built up from the class-warring masses of Shanghai:

> A Japanese soldier patrolled the cinder track nearby. He walked across the grass and stared down at him. Irritated by the noise, he was about to kick him with his ragged boot. But a flash of light filled the [Nantao] stadium, flaring over the stands in the southwest corner of the football field, as if an immense American bomb had exploded somewhere to the northeast of Shanghai. The sentry hesitated, looking over his shoulder as the light behind him grew more intense. It faded within a few seconds, but its pale sheen covered everything within the stadium: the looted furniture in the stands, the cars behind the goal post, the prisoners on the grass. They were sitting on the floor of a furnace heated by a second sun. (285–86)

This "second sun" of nuclear power comes clearly, for Jim as for Ballard, out of the Japanese imperial imagination that so fascinates and doubles the will to grandeur of this hyperactive child of the waning British Empire: "Jim smiled at the Japanese, wishing that he could tell him that the light was a premonition of his death, the sight of his small soul joining the larger soul of the dying world" (286). This will to sublation of the subject into the Empire comes out of the Japanese ideology of the modern, as Naoki Sakai has argued in studies on the Japanese reworking of German epistemology to serve the "Empire of the Sun."[13] In Ballard's uncanny complex, the nuclear bomb enacts a fascination with solar energy as an imperial symbol of self-divinity; the bomb sublimates the human into the divine in a deadly body-transfiguring way, a scenario foreshadowed by the kamikaze pilot with the zero/sun on the wings of his plane refusing to die in a Shanghai swamp. The British boy is fascinated by the Japanese, even roots for their victory, because their martial spirit is so grand and supreme in its discipline and self-overcoming. Within Jim's imagination, as it tries to mirror and double the martial imagination of the kamikaze pilot, the rising sun of Japan constitutes an icon of self-abnegating sublimity as much as one of the emperor's divinity. The Americans, with their gigantic B-29s and infinite tins of Spam, enter to fulfill the same imaginary role as heroes of the solar sublime.[14]

Ballard's *Empire of the Sun* helps to image forth, with uncanny mass-media accuracy, that this world war comes deeply out of the modernist fascination with technocratic supremacy and the grand overcoming of nature. As Jim remarks at the outset of the war in Shanghai, "But the bombardment of the *Petrel*, the tank that had crushed the Packard, the huge guns of the *Idzumo*, all belonged to a make-believe realm. He almost expected Yang [his Chinese chauffeur] to saunter into the ward and tell him that they were part of a Technicolor epic being staged at the Shanghai film studios" (46). This technocratic supremacy on an epic scale is part of what fascinates Jim about the Japanese: "Aircraft had always interested Jim, and especially the Japanese bombers that had devastated the Nantao and Hongkew districts of Shanghai in 1937. Street after street of Chinese tenements had been leveled to the dust, and in the Avenue Edward VII a single bomb had killed a thousand people, more than any other bomb in the history of warfare" (19).

Jim's shifting allegiance of hero worship from the Japanese to the Americans is partly instigated by the iconic sight of the B-29s flying overhead more than by the bumbling, crass, piratical Basie: "The B-29s awed Jim. The huge, streamlined bombers summed up all the power and grace of America. . . . What impressed him so much was that these complex machines were flown by men such as Cohen and Tiptee and Dainty. That was America" (236). Mingling the vulgar and the sublime into a technocratic icon, the American B-29s sublate the Japanese Zeros whose will to self-abnegation cannot defeat the will to power of the American weaponry, emerging here on a new scale of grandeur and sublime significance: "Two of its [Superfortress] engines were on fire, but the sight of the immense bomber, with its high, curving tail convinced Jim that the Japanese had lost the war" (236). As citizens enter the postmodern condition, the ground of war can become an image of global transformation.

As Ballard notes in his wry foreword to the novel, "The military airfield at Hugngjao is now the site of the Shanghai International Airport."

By gathering documents that support the postwar creation of a "white paper" of atomic victims, shame, misery, and radioactive wounding, Ōe's *Hiroshima Notes* (1965) helps to rectify any American fascination with the sublime power and earth-shattering danger to humanity and the future released at Hiroshima. Documenting atomic suffering, Ōe does acknowledge that "these awesome [nuclear] weapons reign over our age like raving-mad gods," such as when Nikita Krushchev boasted to John F. Kennedy and the world that the Soviet Union possessed "a fearful weapon capable of exterminating mankind."[15] Ōe focuses, instead, on drawing human parables from the specific and relentless suffering endured by these sublime weapons inflicted on the common people of Hiroshima, who are known for their taciturn fortitude.

While nuclear power at times of spectacle enraptures as icon of power, "sublimity" is by no means the whole issue or effect in representing Hiroshima. As Toshihiro Kanai, an editorial writer for *Chugoku Shinbun* of Hiroshima, wonders about the political rhetoric of the antinuclear movements emerging in the 1960s: "Is the atomic bomb known better for its immense power or for the human misery it causes?" (67). The assumption is that any such fascination with the sublimity ("immense power") of the atomic bomb need not repress the misery inflicted on the people of Japan, whose strength and moral fortitude is called on to overcome even that disaster. This radioactive suffering must not be forgotten, and, in Ōe's existential calculus, it can lead beyond catastrophe and tragic absurdity to inform the basis of an antinuclear resistance rooted in "Hiroshima" as fact and image.

Although *Hiroshima Notes* is far more acute geopolitically than John Hersey's *Hiroshima*, Ōe's focus remains existential, stressing the emergence of individual "courage in the face of desperate anxiety"; individual dignity; and the undefeated humanist decency of doctors, journalists, and housewives (53). Amid the ruins of liberal modernity he searches for "an authentic man" like Doctor Shigeto and, everywhere, "the dignity of man." As in *A Personal Matter* (1964), which frames the personal tragedy of Bird's deformed child against a nuclear protest movement that does not so much diminish as exemplify acts of human resistance, the will to suicidal despair and self-pity must be countered by a will to political action linking the personal to the communal.[16] In *Hiroshima Notes*, the cry "No More Hiroshimas" links the horror of personal wounds and the taciturn morality of Hiroshima victims to the communal will to hope, resist, and build up global peace across the postwar world. Remembering Hiroshima, and forgetting world war, the need abides "to create a new style of peace movement centered in Hiroshima" (37).

Repression of Hiroshima comes naturally to the national rhetoric of neoliberal and market-worshiping Americans, however, transcending a history ever their own. As Ōe observes, in 1945 the US Army Surgeons Investigation Team claimed that all people expected to die from radioactive effects had already died and that no further physiological effects due to residual radiation would be acknowledged (60). This American statement proved to be obscene, as citizens of Hiroshima and Nagasaki went on suffering and dying from keloid wounds and tubercular effects in monstrous ways the "white paper" sought to document. Ostracism conspired with shame to make the situation of the sufferers even worse. Occupation restrictions on Japanese journalism enforced repression, as the wounded were, in effect, urged to keep silent and pass away quietly (66). During the Tokyo Olympics of 1964, when a young man born in Hiroshima on A-bomb day was selected as the last runner to carry the Olympic flame, an American journalist protested that this was an unhappy choice because it reminded Americans of the atomic bomb (99–100). This journalist, as Ōe shrewdly perceives, "preferred to erase all traces of Hiroshima from the American memory" (100). Representative Americans, remaining Emersonians by being "endless seeker[s] with no past at [their] back," would rather forget history and remember the techno-euphoric future of friction-free global capitalism, but only if the globe takes shape after their own national dreams.

The people of Hiroshima cannot forget Hiroshima, however, even though "they have had enough of [remembering] Hiroshima" (101). As Ōe says, within a Cold War context of historical repression and the sublime idealization of nuclear force: "In this age of nuclear weapons, when their power gets more attention than the misery they cause, and when human events increasingly revolve around their production and proliferation, what must we Japanese try to remember?" (90). What must Americans do to remember Hiroshima and help to usher in a nuclear-free future disenchanted with the project of technological domination? Living in an era of nuclear proliferation and the technocratic deformation of nature, we are all, in some sense, what the Japanese term *hibakusha*—literally, "explosion-affected people." As a rupturing point in the march of historical progress, Hiroshima deforms an American horizon of sublime political agency.[17]

Hiroshima puts an end to ego-centered modernist imaginings of the nation. Though we may not have directly suffered the radioactive effects within two thousand meters and fourteen days of the bomb's hypo-center at Hiroshima, Hiroshima haunts, fascinates, and inhabits us as recurring memory, as trauma, instigating a collective labor of mourning, as Marguerite Duras's screenplay *Hiroshima Mon Amour* memorably portrays.[18]

This to some extent explains the mythic framework of geopolitical and subjective mourning as soul making invoked by Michael Perlman in *Imaginal Memory and the Place of Hiroshima* (1988) to underwrite the claim that "lasting [nuclear] impressions do implicate the (largely unconscious) ways in which we are all survivors of Hiroshima, and hence have to some extent incorporated, identified with, and internalized the sufferings of actual *hibakusha*."[19] Using representations of Hiroshima as *imagines agentes* of the political unconscious, Perlman labors to house "Hiroshima," as catastrophic place and mythic metaphor, in the postmodern memory working to evaporate history and to forget. It is Perlman's utopic plea, grounded in the unconscious power of archetypal mourning, that "images arising from the place of Hiroshima speak to us: Remember Hiroshima" (vii).

In *Thank God for the Atom Bomb* (1988), building on a polemic he wrote in 1981 titled "Hiroshima: A Soldier's View," Paul Fussell enacts another mode of "remembering Hiroshima": the atomic bomb recalls not an American memory filled with shame, trauma, or guilt but one of pride, honor, the kill-or-be-killed war. The fascinating sublimity of the atomic bomb has proved redemptive within history and need not occasion moral ambiguity, retrospective shock, or shame within formations of postmodern culture. "The degree to which Americans register shock and extraordinary shame about the Hiroshima bomb correlates closely with lack of information about the Pacific war," Fussell suggests.[20] American history need not shame or chasten but empower, vindicate, self-forgive. The Hiroshima sublime, likewise, need not induce blockage, negation, terror. Fussell's blood-and-guts title is not at all ironic. He does thank God for putting the atomic bomb in the keeping of exceptional Americans and thereby ending "the unspeakable savagery of the Pacific war," if not installing America at the historical center of such technocratic power.[21]

Americans such as John Kenneth Galbraith and Michael Sherry, who argue against the bomb and theorize its image and consequences retrospectively, according to Fussell, are suffering from "a remoteness from experience" or a "rationalistic abstraction from actuality" (19–20). The atomic bomb debate can be "reduced to a collision between experience and theory."[22] Like Walt Whitman, Fussell knows: he suffered and was there, though not actually at Hiroshima (he might then have had another view of the morality of the bomb) but serving as an infantry lieutenant about to invade Honshu. Defending the bomb's dropping—though worrying about and postponing, later in the essay, the nuclear consequences for ecology at places such as Three Mile Island—Fussell opposes abstract theory as "error occasioned by remoteness from experience" to the material "truth by experience" of dog soldiers like himself and President Truman, whom the atom bombs dropped on Hiroshima and Nagasaki spared from infantry invasions of Japan, hence saving countless American lives.[23]

In Fussell's moral calculus of American power, the bomb had a redemptive and egalitarian function on Japan, making this once ruthless nation for which "universal national kamikaze was the point" into a potentially democratic, constitutional, law-abiding country of citizens like ourselves: "It is easy to forget, or not to know, what Japan was like before it was first destroyed, and then humiliated, tamed, and constitutionalized by the West."[24] Fussell's syntax is embarrassing yet exact: Japan is "first destroyed" only later to be subjugated and redeemed by the discourse of the West. The atom bomb had done its sacred American labor of regeneration through violence. America remains an exceptional country for which the greatest violence can somehow still do the most good. In effect, the sublimity of the bomb had redeemed Japan from itself and opened up Asia Pacific to American postwar hegemony.

Those Americans who emote about the horrors/wonders of the atomic bomb need not "dilate on the special wickedness of the A-bomb-droppers."[25] A dehumanizing rhetoric was called for. Justifying the wartime ritual of hanging Japanese skulls on bayonets, Fussell must concede the Orientalist premises of his will to sublation of Japanese otherness into animal-like opponents of the American sublime: "Among Americans it was widely held that the Japanese were really subhuman, little yellow beasts, and popular imagery depicted them as lice, rats, bats, vipers, dogs, and monkeys."[26] Fussell *knows war by experience*; he fought, bombed, and killed, yet he has no trouble sleeping at night or remembering world war in the Pacific and justifying Hiroshima. Only remote-from-experience postwar theorists can still linger over the horrors, ambiguities, and death-inducing terrors posed by the nuclear sublime.

That the experience of common citizens might lead one automatically to justify the goodness and rightness of the atomic bomb might come as a horrid surprise to Japanese novelists such as Ōe and Masuji Ibuse, whose *Black Rain* (1969) uses a species of quiet realism and journal-based documentation to try to establish the horrid effects of the bomb on noncombatant citizens. The narrator, Shigematsu, is opposed to any flights of moral or political theorization and keeps the journal he is reconstructing as close to the actual experience of the bomb's dropping and aftermath as he can:

> "Well," he said, "I got through a lot today. I've copied it all out up to the place where the West Parade Ground is jammed with people taking refuge from the mushroom cloud. Even so, I haven't got down on paper one-thousandth part of all the things I actually saw. It's no easy matter to put something down in writing." His wife counters, "I expect it's because when you write you're too eager to work in your own theories," but Shigematsu responds, "It's nothing to do with theories. From a literary point of view, the way I describe things is the crudest kind of realism. By the way, have these loach been kept in clean water long enough to get rid of the muddy taste?"[27]

This activity of cleaning and replenishing the fish and carp ponds of Hiroshima is just one of the little rituals Shigematsu enacts to help nature and humanity recover from the bomb. The journal, too, becomes an act of historical cleansing. Yasuko, his niece, cannot forget the black acid rain that marred her as a Hiroshima survivor; others in Japan, refusing to marry her, will remember for her. Nature and ordinary custom, everywhere in Hiroshima, have been irrevocably disturbed: "The bomb seemed to have encouraged the growth of plants and flies at the same time that it put a stop to human life."[28] Years later, the Blessing of the Dead Insects by the farmers of Hiroshima takes on an odd, surreal significance, recalling the deformed, maimed, blasted human corpses who had died like insects.

The NBC movie *Hiroshima: Out of the Ashes*, broadcast on August 5, 1990, provided at least one American attempt to move toward representing/remembering Hiroshima and thereby activating a collective process of mourning some forty-five years after the event. It is melodramatic in scale, allowing a politics of identification to take place in which postwar Americans identify not with the technocratic sublimity of the bomb but with its human victims. Arguing that the made-for-TV movie offered "an unflinching look at the physical and emotional devastation" that was wrought by the American atomic bombings, the critic Christopher Hull yet had to demur: "There is something disturbing about the fact that it has taken this long for the story about the bomb's victims (more than 100,000 were killed in Hiroshima) to be dramatized on the small screen." As a narrative of this sublime event, the movie drew on humanistic stereotypes from John Hersey's *Hiroshima* (Father Siemes is here the German Jesuit priest spiritually transformed by the bomb to a stance of compassion toward ordinary Japanese), Ballard's *Empire of the Sun* (two boyish American soldiers wander miserably around Hiroshima after being freed by the bomb), and Ibuse's *Black Rain* (the stoic Japanese doctor is Dr. Hara who moves away from being a war advocate toward unflinching labor, and the actor Pat Morita plays a postman who holds on to small Japanese rites amid the devastation) to illustrate, as Hull puts it, "the resilience of the human spirit."[29]

Filtered through this intertextual network, the bomb is remembered by Americans, in other words, to summon their basic humanity, which this primal scene of death and destruction seems to contradict. Creature of technocratic sublimity and apocalyptic doom, both myth and monster, the Hiroshima bomb had instigated an ongoing sense of terror for scientist, doctor, soldier, and citizen alike. Japanese filmmaker Akira Kurosawa's visionary *Akira Kurosawa's Dreams* (1990) takes a drastic approach to this aesthetic/ethical mandate of remembering Hiroshima and portraying the threat of the nuclear sublime to nature's ecosphere. In the "Mt. Fuji in Red" sequence of this subliminal narrative, five nuclear reactors explode and spontaneously devastate the landscape of contemporary Japan, destroy its cherished icon of natural sublimity, and drive thousands of its citizens to flee into the radioactive sea. A flower-laden Japan is turned into an arid purgatory of ash, debris, and ruin. In the next sequence, "A Weeping Demon," Kurosawa tries to imagine a postnuclear Japan, with monster dandelions and cannibalistic demons who wander around the volcanic landscape of blood and nuclear ash as in some Buddhistic hell.

Even death seems contaminated, and the afterlife itself disturbed, by the nuclear horrors twentieth-century humanity has wrought in its technocratic overreaching. Having evoked this apocalyptic dynamic of the nuclear sublime, Kurosawa can only retreat from this postmodern abyss of radioactive contamination into a premodern landscape of rural Japan, evoking a "Village of the Watermills" where old and young cherish the vitality and path of nature and live by customs and rituals that provide happiness and a community long vanished from post–Hiroshima Japan. The mighty Asian/Pacific consumer culture linking America and Japan can induce post-historical forgetting of the wartime Pacific, normalizing the trauma of the Hiroshima sublime into a vast Asia-Pacific Economic Cooperation (APEC) market for Asian and European investment, beyond ideology as it were. As Ōe claims at the end of *Hiroshima Notes*, "Outside Hiroshima we are able to forget the misery in that city; and forgetting has become easier with the passing years, now twenty [forty-five], since the atomic bombing" (140).

Even inside Hiroshima, these transnationalizing days of global post-modernity and the dismantled poetics of region and place, one can forget "Hiroshima," go to Hiroshima Carp games, and buy the latest designer fashions from around the Earth. Americans and Japanese can ponder the A-Bomb Dome, which lingers in Hiroshima as the first blasted allegory of postmodern art. Formerly the Industrial Promotion Hall of Hiroshima Prefecture designed by the Czech architect Jan Letzel and built in 1915 as one of Hiroshima's first modern buildings, it is now a permanent memorial. On July 11, 1966, despite escalating real estate costs and an emerging will to forget, the Hiroshima municipal assembly voted in favor of permanently preserving the building's destroyed dome, a blasted icon of modern engineering torn open, deconstructed, and deformed beneath the sky for all to ponder and to *remember*.

The deformed dome at Hiroshima remains not so much a US Pacific icon dedicated to the sublime power of the bomb as, Shinoe Shōda claims in her collection of post-Hiroshima poems *Penitence*, "a monument dedicated to peace for all mankind."[30] Moving beyond the agon of national superiority and nuclear might in the troubled yet promissory waters of the Pacific, where unstable political powers such as North Korea and Russia still threaten to use nuclear weapons across the region, let us learn, in "worlding" forms of spiritual quest and risk, to make new alliances of peace and synergistic creativity reign across Asia Pacific and the shared ocean.

SEVEN

Waking to Global Capitalism and Oceanic Decentering

Reworlding US Poetics across Native Hawai'i and the Pacific Rim

Discussing cities is like talking about the knots in a net: they're crucial, but they're only one part of the larger story of the net and what it's supposed to do. It makes little sense to talk about knots in isolation when it's the [trans-Pacific] net that matters.

· KIM STANLEY ROBINSON, "Empty Half the Earth of Its Humans"

In this chapter, I will again pressurize contradictory meanings of "Pacific Rim" as a cultural-production framework and transpacific nexus, or what has been troped as translocal "net" (as Kim Stanley Robinson situates such digitalized vanguard cities) more than to forms of Euro-American or Asian "modernism" as such, which others (from Fredric Jameson's temporal mapping of capitalist totalities to Evelyn Ch'ien's modes of mongrelized modernity in *Weird English*) have already done in speculative ways.[1] As a contributor to creative writing and cultural studies forces of "minority becoming" in sites such as Honolulu and Seoul across the region, I have figured forth an "Oceania" as set within and against the hegemonic frameworks of the Pacific Rim and Asia Pacific as transnational regions of area studies. In this regard, I published a global-local Tinfish Press pamphlet—or small press manifesto of sorts—entitled *Pacific Postmodern*, wherein the paradoxical mix of place-based and indigenous longings in Hawai'i, as across Oceania, clash against a will to pidginized code-switching and multilingual postmodern experimentation.[2] When Marjorie Perloff, the comparative literature scholar of postmodern poetics, read a draft of this work, she wrote an email back to caution against easy embrace of an affirmative localism with the tender rejoinder in her subject heading, "Pacific Most Modern." She was suggesting (perhaps rightly as a formal matter) that this kind of Pacific experimental work was not all that experimental or postmodern in effect and was still just Pacific Most Modern in its style, forms, and genres. As for the place-based claims to represent voices of cultural authenticity, "'nuff said already," as Da Pidgin Guerilla Lee Tonouchi would say in his in-your-face use of Hawaiian pidgin English, which I pushed him toward as a poetry student in my English 313 undergraduate course at the University of Hawai'i, Mānoa in the 1980s. In those writing classes I advocated for pidgin works of local imagination by Eric Chock, Joseph Puna Balaz, Michael McPherson, Wayne Westlake, Diane Kahanu ("Hey, Just 'Cause I Speak Pidgin No Mean I Dumb" worked well in class study), and others as examples to emulate in style, voice, and form and to express an island-world ethos.[3]

As a signifier advancing techno-urban development, the Pacific Rim still enchants and hails (*interpellates* is the Althusserian verb for this calling to subject position) diverse enframers in its mysterious allure, as if this geoimaginary gestured toward a kind of utopic global destiny calling disparate countries from Korea and Taiwan to Chile and Australia into its *transnational* coalition of region, place, community, motivation, and cultural identity.[4] It is as if the vast Pacific Ocean and its alluring ring of volcanic formations still configures and links this "Rim" into a common planetary future and becoming-capitalist urban formation. Islands in such a calculus hardly signal any strong difference from the mainland even as they are (by archipelagic ties) attached as peripheral appendices, whether Taiwan to Mainland China, Okinawa to Honshu Japan, Cheju Island to the domination of the capital city of Seoul, or Hawai'i to the continental United States. Still, to invoke Balaz's pidgin-voiced and indigenous-affiliated countergeographical affirmation, "Hawai'i is da mainland to me."[5]

With subterranean wit, political imagination, and active joy, the poetry in Balaz's thirty-five-year comprehensive collection *Pidgin Eye* (2019) presents, in its multivoice pidginized plenitude, an uncanny reclaiming of Hawai'i as people and place in the local oceanic Pacific called Oceania. In a fabulously pedagogical playhouse of voices, Balaz tracks as well as mimes an "anti-paradise" of contemporary Hawai'i Nei filled with pit bulls, skateboarders, University of Hawai'i Bows fans, drag queens, bullies, crazed postal workers, gamblers, canoe girls, Rastafarians, local sex-hungry gangsters, gods and goddesses speaking the language of the Pearl Ridge mall, hula profiteers, losers, cheaters, lost souls, and sacred seekers, as well as Hawaiian Sovereignty advocates speaking in a passionate if meandering style. Balaz writes—over all these disruptive years and postmodern changes, on and about his home island of O'ahu since the late 1970s, editing the journal *Ramrod* and making the first anthology of contemporary Hawaiian poetry written in English—to create a living playhouse of multiple languages, street attitudes, and colorful styles. He uses the expressive humor of the mixed, impure, impious, submerged, and recalcitrantly new in such ways that he makes the voice of William S. Merwin from Haiku, Maui, seem at times too canonical and romanticized. As with the "Calibanization" of English and nation-language activated by the Caribbean poet Edward Kamau Brathwaite, Balaz's pidgin of Hawaiian Creole English (preserving strong ties to Hawaiian language and values more so than to standard, or "haolified," English) becomes a subversive means (as in the raunchy love poem, "Lapa Poi Boy") by which "English is not so much broken, as broken into" by the always moving and subversively punning *Pidgin Eye/I/Ay*.[6]

Falling Off the Pacific Rim: Becoming Rip Van Winkle

In the fall of 2004, I returned to South Korea as visiting professor at Korea National University of the Arts to work on understanding, in later phases, these amplified inter-Asian and Pacific cultural modes of globalization; Pacific Rim visions, major and minor; and "waking to global capitalism" in its expansive and richly innovative modes. This second-time-around visit became reeducation in the disappearance of aura and the shock of a perpetual urban modernity, altering all that I had taken as poetic orientation points of everyday existence and urban survival when I was working and living in the capital of South Korea from 1982 to 1984. On the Asianizing Rim this time around I felt less like a clunky Texan giant stepping out of the television set than like a shrinking American ghost or neoliberal Rip Van Winkle searching (along the nuclearized Demilitarized Zone border) for some past space or everyday modes of cultural interaction that by then had all but vanished into globalization as hybridization.

Eight or nine Korean baseball players are now playing in the US Major Leagues. By wireless online in Seoul I could read the *Newsweek* cover story about Bob Dylan's *Chronicles* before a Boston friend (Lindsay Waters at Harvard University Press) had learned that the unexpected autobiography was coming out. Globalized in mores and modes, as well as linked to not one but two subway stations in a revived neighborhood in eastern Seoul, Korea University had gone from being "Makoli Dae-hakyo" (nicknamed for a rural Korean fermented rice drink) to having a glassed-in Starbucks on campus and declaring its campus "centennial 'memorial'" drink to be a Chateau La Cardonne wine from the Medoc region of France. At times, I felt like one of those incoming country bumpkin students from rural Korea that Korea University was so proud to educate, across the modernizing twentieth century, and yet I was coming there to live and teach from the so-called vanguard Northern California region on the Rim.[7]

The Asia/Pacific Rim space of Seoul goes on becoming a "technoscape" of cell phones, wireless communications, bullet trains, cell phones, flat screens, and digital interface; it is not so much *behind* the globalizing West (California North and South has yet to install even one bullet train), that is to say, but *up ahead* as global-local urban form linking city and country in a space-time globalized maze of libidinal interface on the Pacific Rim.[8] Its K-Pop groups and romantic comedies on Netflix reign from San Francisco and Honolulu to Paris. It is not enough to figure these modes of cultural production in Asia or on the Rim as the strategies and technologies of perpetual latecomers or late-modern derivatives, leapfrogging strangers, or bastardized mixtures in any sense. The postcolonial Empire that is the United States, combined with the New Europe, has much to learn from a pedagogical immersion in the everyday mores and modes of Pacific Rim global cities such as Seoul, Shanghai, Tokyo, Sydney, Jakarta, and Singapore, if not Hong Kong.

Rip Van Winkle's claim, "Strange names were over the doors—strange faces at the windows—everything was strange," speaks to an early trace of this perpetual capitalist-modern disruption and *deworlding*.[9] This world-decentering and uncanny displacement of homeland spaces and cities is being affected by un-convivial strangers from abroad (such as my ancestors from Scotland, Ireland, and Italy), back when (to invoke the Asia-ignoring postmodernist Jean Baudrillard) "America [was] the original of modernity and Europe is the dubbed or subtitled version."[10] Transpacific modes of these immigrant displacements and global-local interfaces continue to make the United States at times look like Van Winkle's Hudson River hamlet and less like that venture capital-rich yet ideologically sleepy Silicon Valley suburban haven—namely, Palo Alto, California.

This Pacific Rim consciousness of leadership, vanguard cultural semiotics, and techno-innovation is not, in these more traumatized post-9/11 contexts and the global pandemic across inter-Asia and the world, a generally available mode of world reckoning. Indeed, the Pacific Rim may be just another dying or aborted imaginary of neoliberal global or "worlding" conviction, weakening even "in its hold [over] the US global imaginary" since the collapse of the socialist bloc and the seeming demise of Cold War civilizational binaries, even as this strategic region is given a new urgency since the days of George W. Bush regime's turn back to security-state binaries and borders.[11] Within these sites of Rim-based cultural production, as in the journal *Inter-Asian Cultural Studies* and pop-culture genres such as manga and the telenovela, the turn to inter-Asia regionalism that amplified during the 1990s in many ways means, in effect, a hugely consequential turn back to China and Japan as geomaterial-*cum*-cultural core.

A "return to Asia" and "inter-Asia" as a method of localized cross-referencing implies, as well, a regional consolidation into Asian transnational solidarity that regrettably very much occludes and ignores the interior Pacific and sublimates the lurking heritages of Japanese imperialism across Asia, not to mention the staunch differences of Islamic Asia across the region from Manila to Jakarta and Mumbai. This "return to Asia" also presupposes a form of region making that is all but completely non-oceanic in its spatial and cultural turn back toward some continental inter-Asia that is China-centered and, at times bitterly, reactive and reactionary in relation to the United States and the white nationalism of the West as those "running imperial dogs" and so on. China's cartography of dominated frontier toward Tibet, or towards the "renegade" but internationally recognized island state of Taiwan, gets very much washed out in this sublimating turn back to "harmonious" ex-imperial China that wants some global respect and the infrastructure of regional autonomy and one-party control with "Chinese" characteristics, heightened surveillance, and minimal political democratization if ever.

Still, despite its array of historical and cultural-political contradictions leading the region from Chairman Mao to Andy Warhol and back, we might want to hold on to the Pacific Rim imaginary in all its transcultural and transnational dimensionality as something worth linking to and preserving, if utopic and visionary in its decentered, translocal, and interlinked regional vision of Rim cultures as a transnational and planetary solidarity movement. Conceiving the Pacific Rim as such a transcultural region of imagined transnational belonging across Oceania might allow us to forge a different pedagogy, history, and framework of canonization and museum display, wherein once "minor" or occluded energies and formations are given the renewed agency and hypervisibility they have long deserved as a cultural semiotics.

Transcultural Pacific: Worlding Pacific Islands inside the Rim

Whatever the turn to Asia and return to mainland ex-imperial China, we need to articulate the Pacific becoming Oceania still as a transnationalizing and transcultural space of uneven flows and uncanny or spectral conjunctions with the past. This has been the coalitional project while working in Hawai'i from the late 1970s to 2000 and helps to explain this turn to recalling place-based claims of Queen Liliu'okalani in this chapter. We need to respond to these indigenous, aboriginal, and Native Pacific perspectives on the transnationalizing of the Pacific to articulate how these various "spectral" peoples respond to transformations of the Pacific into regional coalitions of imagined transnational community and confederated solidarities or global ecumene troping "Asia/Pacific" or "inter-Asia" into contemporary forces and global-local peoples. What role can culture or a culture-based politics play in this geopolitical struggle for equality and for resources and recognition? Is culture just doomed to late capitalist subsumption, a hollowing out into what Jameson and other postmodernists call the "Disneyfication" effect of local simulation and deworlding death?

Not to be foreclosed from hope, countervision, or social struggle, we still need to deal not only with land-based claims and national identity issues of an imagined Native Pacific community, but to culturalize the makings of a transnational or transcultural Pacific from multiple angles of Native and neo-Native vision. Can Asia or the dominant US formations hear the Pacific? Can the Pacific enlist Asia into some kind of critical dialogue? Can the interior Pacific be articulated to formations of the Pacific Rim, inter-Asia, or Asia/Pacific without being evaporated in this modernizing and resignifying process of translation? In this contradictory era of internet-fueled democracy leading to the return of authoritarian governance from the United States to the Philippines, how can the mounting "digital divide" and the spread of technological unevenness be overcome? Singapore is wired with more than a million computers, for example, whereas Papua New Guinea abides as everyday Rim culture with less than 1 percent of its computers on the World Wide Web. By many accounts (and most of them *gloomy* as the world market hits another global recession), the Pacific Islands are so entangled in a web of globalization forces that will not let go of the drive to Pacific profits, and many of these forces across Oceania emanate from Asia via Manila, Tokyo, Taipei, Hong Kong, and Singapore. The global tourism industry enlists sites of erotic and primitivist fascination, from Bali to Honolulu and Fiji, even as automobile, garment, biogenetic, and mining industries expand offshore networks across the region.

The globalization agenda of neoliberal market forces manifests little interest in small places or culture-based claims. The second, or cybernetic, wave of globalization, following on earlier colonial patterns, is underway, incorporating these far-flung island territories into the global economy in terms that almost always suit the dominant powers and their bank- and credit-driven institutions, including the International Monetary Fund (IMF), the World Bank, and the Asian Development Bank. Island nations at times appear to be the canary in the coal mine of globalization, crying out in the huge darkness that they are threatened, here and now, with global warming, as well as indigenous, planetary, and ecological extinction of diversity. This is to urge that any strong version of Asia Pacific or inter-Asia, as propagated on the Pacific Rim, cannot afford to forget internal Pacific dynamics across Oceania and claims of precapitalist aboriginality, communal alternatives, and antimodern memory. Asia Pacific and inter-Asia do not just belong to the community of transnational capital or astronaut class of frequent flyers.

We can, in these days of postmodern reason, still seek the antagonistic synergy of postcolonial Asia Pacific forces, flows, linkages, and networks that are spreading across the region. To the post–Cold War United States, Asia all too easily can refract into a space of dynamism, danger, threat, allure, and promise. Nonetheless, we cultural producers in the region of Pacific Rim conjugations have been dreaming of another transnational and transcultural Asia Pacific, not some ocean-submerged continent bespeaking exploration and expansion for marines, politicians, and tourists to claim as site of adventure and metropolitan life writing (i.e., region as "absorbed" into Euro-American geography or Earth writing and world formation).

Given such far-flung global-local transformations, the appeal of the Pacific Rim as a globalizing site of transnational coprosperity and capitalism's transcendental future no longer beckons with quite the same gleaming promise.[12] For a metonymy of this lost-aura effect, consider the graying flat skylines of high-rise Tokyo and the romantic dead-end entropy of loser bars and bad whiskey ads in Sofia Coppola's *Lost in Translation* (2003), or the emperor-worshiping genuflections of Tom Cruise in *The Last Samurai* (2003), which evokes a nostalgic form of Meiji Japan that is so history-denying in its longings that it stands for more sublimated modes of American neo-Orientalism altogether. Any current version of Pacific Rim discourse can no longer just factor in postwar Japan as the sublime dynamo of transpacific transnational *promise* (as model minority in the global sphere) or *peril* (as threat to US-run geopolitical stability and economic supremacy in the region). We must fully factor in the presence of postsocialist Global China on the Rim, with its seismic rural-urban shifts to global capitalist restructuring and huge port city nexus on the transpacific front, as well as multicultural and techno-savvy India of democratic plenitudes, with its booming offshore service resources and modes of data connectivity taking place in its transnationalized megacities of silicon skills that go on overflowing the nation-state confines and Asian regional containment.[13] North American ambivalence toward Asia (its promise as mimic model as well as peril as rising threat) is a function not only of deep national psychic structures of racialized difference but also of geopolitical entanglement in the Pacific, nation, place, and world.

The trans-binary Pacific Rim effect of imagined transnational community can still function as a key worlding space of global flows suggestive of an emergent, as-yet-untheorized formation we have been gesturing toward as Oceania. Whatever the stasis and dead-end effect of global capitalism as portrayed in the burned-out careers, lost romances, and entropic Tokyo urbanscape in global slacker movies such as *Lost in Translation*, the US Pacific Rim remains a crucial locus for global capitalist dynamism in all its hyperspeculative risk, as well as more phobic labor-class modes of transnational othering and transcultural becoming. Admittedly, this Pacific Rim nexus remains a swirling and uneven formation, full of huge rural-urban imbalances and labor-capital injustices and IMF ruses, aggravated by contradictions of peril and promise after 9/11 and the global pandemic of 2020 and 2021.

Whatever the peril and promise the Pacific Rim poses as a region of transnational interface and transcultural possibility, we need to expose the discourse of American Orientalism as a neoracialized process of geopolitical othering that enacts what Gayatri Spivak calls the "crisis manage-ment" of transnational circuits and planetary emergence.[14] This phobic othering persists at the same time the Asia Pacific region of Asia-Pacific Economic Cooperation (APEC) is morphing into a porous, user-friendly, globalizing space of post–Cold War, post-binary, "post-Orientalist" in-teraction and communal hybridity—*the* neoliberal capitalist space, as it were, of global-transnational flows.[15] As the economy continues to grow across the Asia Pacific region, Arif Dirlik has suggested, the Pacific as a geopolitical space needs to be redefined so that its future might belong to the peoples in Asia Pacific as across Oceania.[16]

Along with Chalmers Johnson, whose anti-Empire work includes *Blowback* (2000) and *The Sorrows of Empire* (2004), wherein the Asian Rim figures as a security region, Meredith Woo-Cumings warns that the Pacific still serves as an American lake, in the sense that the United States "holds sway not just with the Seventh Fleet, myriad military bases, and a panoply of high-tech and nuclear weapons, but also with a U.S. presence that in all its forms—cultural, political, economic—remains pervasive." Woo-Cumings, following a structural stress on military and economic hegemony, misses the booming assertiveness of cultural power in sites such as Singapore and Seoul or Kyoto when she claims, "East Asia lacks the language and psychology for self-assertion, which is an artifact of its long domination by the West."[17] Who can believe such large-brush claims anymore for cultural or racial abjection on the Pacific Rim or inside Oce-ania? We need to hear the return of voices from the Hawaiian sovereignty struggle for the indigenous unity of mountain and ocean, for breathing room and taro fields, for water and air, and for a beneficent ruler whose poetry and song were modes of history-making.

Although memories of war, racism, and empire haunt the postcolo-nial region we live in as "inter-Asia" or "Asia/Pacific," we can retrieve a different regionalism and thus hope to counter (via alternative coalitions, creations, theories, and movements) any lurking expansionist designs or proto-colonial dreams of conquest or containment in this, our shared transnational Pacific Rim region. Transitioning to a new era takes much imaginative work, for distancing the present, evoking other pasts, and conjuring alternative futures across the Pacific or Wansalwara (one salt-water), as a pidgin English term from Papua New Guinea riffs on oceans of Oceania. This turn to a pidgin-speaking Oceania seems a Queequeg-like turn to counter Captain Ahab's ocean-engulfing hunger for sacral-ized commodities in American Pacific desires of *Moby-Dick*.

Korean films can allow us to see a Pacific Rim global city where activism has not died out but has gone underground in ways (as in such class-haunted works as Pak Chan-uk's *Sympathy for Mr. Vengeance* [2005] and *Oldboy* [2005], as well as Bong Joon-ho's *Snowpiercer* [2013] and *Parasite* [2019]) that demand Anthropocene attention. The world of cultural production is being "recentered" in ways that recall the haunting of history and deformations we cannot forget in our drive for the internet, along with debt systems of capitalism and its global popular imaginary of life in *Crazy Rich Asians* (2018) and in the endless yoga seminars and TED talks across life in Zoomlandia as a form of digitalized "worlding" that some say is here to stay.

Reworlding Naboth's Stolen Vineyard in Hawai'i

All of modern American literature in Hawai'i, from a Pacific Islander point of view, as situated within decolonizing politics and primordial worlding tactics taking place across Oceania can be said to begin with *Hawaii's Story by Hawaii's Queen*. This work is a haunted memoir written (from late 1896 through 1897) by the deposed Hawaiian monarch Queen Lydia Lili'uokalani, whose reign underwent the tragic downfall of the Hawaiian Republic. Published in Boston in 1898 amid the imperial scramble by what she called the "Mana Nui" (Great Powers) across the Pacific, as across Africa, the memoir contests this overthrow of a legitimately constituted sovereign nation by an array of white-settler forces (mostly Americans from New England and California) in 1893. The memoir, read rightly as an oceanic and islander text of world-making importance, activates the power of culture and implicit meanings of indigenous poetry to serve as a *countermemory* to settler-colonial US Pacific hegemony.

Testimony, photograph, anecdote, diary, newspaper item, and treaty, narrative, and lyric commingle in *Hawaii's Story* into a Christian Hawaiian appeal for justice and international recognition from an imperializing American government that, during the William McKinley administration, was engaged in extraterritorial expansion into Asia and across the Pacific Ocean, as well as in the Caribbean archipelago.[18] Even as she worked to compile her well-known songbook *He Buke Mele Hawaii*, after her imprisonment for eight months in Iolani Palace in downtown Honolulu, Queen Lili'uokalani, the last Hawaiian monarch, was accused of adultery, sorcery, and tyranny and lampooned in print and cartoon as "a portly, chocolate colored lady," as one journalist put it in *Leslie's Illustrated Weekly*. She was also declared (more accurately) in Boston papers "a devout and perfect Christian" lady after she had contributed to a charity show a Hawaiian doll, in a Mother Hubbard gown she had handsewn, "a very pretty doll, that resembled somewhat some of my people who had intermarried with the foreigners."[19]

As a skilled *haku mele*, this "poet-maker" of more than two hundred songs—including the first Hawaiian song to be published on the American mainland ("Nani Nā Pua" [1869]) and the first Hawaiian national anthem ("He Mele Lāhui Hawai'i" [1866]), during the reign of King Kamehameha V—composed many works of song lyrics filled with what she called in one of her *mele inoa* (name chants) "a troubled aloha."[20] "Aloha 'Oe" (Farewell to Thee) is not just a song about lovers parting; it is also coded with the layered pathos of separation, brokenness; it is a hello and farewell to Hawaiian nationhood as such. Queen Lili'uokalani's works— as a model for future activism for Native Hawaiians—have influenced Hawaiian-based authors, from the singer-activist George Helm of the demilitarizing struggles and of the 1970s and early 1980s to the novelist and playwright descendants of this Hawaiian and Oceania vision, John Dominis Holt, Victoria Kneubuhl, Joe Balaz, Alani Apio, and others.

Writing against such national dispossession, Queen Lili'uokalani laments in her memoir and related poems that this is a consequence of the "grand Monroe doctrine" (372) invoked against the Spanish Empire, as it earlier had been (in Hawai'i in 1842) against British and French national expansion into the Hawaiian Islands. Six months after *Hawaii's Story* was published, to no avail, the United States would ratify the annexation of Hawai'i on July 7, 1898; in 1959, it would become the fiftieth state. The 1897 diary that contains drafts of the book would be misplaced, lost, or (most likely) confiscated. Skeptical reviewers would discredit *Hawaii's Story* as "ghost written by Boston journalist Julius A. Palmer," as would the missionary offspring and newspaper heir Lorrin A. Thurston in the 1930s, but the charge is spurious. The queen's memoir haunts American literature with its history of injustice and Native dispossession, as in the US Apology Resolution to Native Hawaiians made into law in 1993.

Queen Lili'uokalani, placed under house arrest in 1895 and 1896, had turned to the translation into English of *Kumulipo*, the Hawaiian creation chant, and to composing lyrics to communicate obliquely with her people. As she notes in her memoir, "The ancient bards of the Hawaiian people thus gave to history their poems and chants; and the custom is no different to this day and serves to show the great fondness and aptness of our nation to poetry and song" (53). Writing in diaries in 1896–97, she keeps Hawaiian selfhood and nation alive, if only as her "story" that would endure shocks of American modernity. Dispossession of Native discourse is the situation Queen Liliʻuokalani exposes, telling "Hawaii's story" as plight of colonial subjection and sacrifice, in the face of Washington's "project to annex Hawaii to the American Union" (38).

Identifying her islands with "the vineyards of Naboth" in the land of Jezreel and the United States (as Melville's jeremiad novel in the whaling-era Pacific, *Moby-Dick*, had warned) with "the punishment of [the Samarian King] Ahab," from I Kings 21, the queen appeals to her American readership across ethnonational binaries of an imperial cartography dividing "us" from "them," white people from Black people, metropolitan powers from peripheral nations, and even so-called islander heathens from converts such as herself:

> Oh honest Americans, as Christians hear me for my downtrodden people! Their form of government [constitutional monarchy] is as dear to them as yours is precious to you. Quite as warmly as you love your country, so they love theirs. With all your goodly possessions, covering a territory so immense that there yet remain parts unexplored, possessing islands that, although near at hand, had to be neutral grounds in time of war, do not covet the little vineyard of Naboth's, so far from your shores, lest the punishment of Ahab fall upon you, if not in your day, in that of your children, for "be not deceived, God is not mocked." (373)

In a transpacific US country that prides itself on the amnesia of its exceptionalism from colonial aims, we still need to recall Liliʻuokalani's uncanny trope recalling Ahab's punishment and lament for her injured "aboriginal people": "Is the American Republic of States to degenerate, and become a colonizer and a land-grabber?" (372).

Lydia Kamakaʻeha (1838–1937), a sister of King David Kalākaua, had married the American merchant John Owen Dominis in 1862 and succeeded her brother to the throne in 1891 after decades of "tormented aloha," when the legitimacy of this royal lineage, as well as tactics of Native insignia and display, was subjected to mockery by Native forces and foreign settlers in the kingdom. Lydia was learned and well-traveled and had attended Queen Victoria's Golden Jubilee in 1887 in London, where she was warmly received, as she was later to be in the salons of Boston and the corridors of Washington, DC. Her genres of literary resistance to colonial history were cultural and poetic, as befit prior Hawaiian traditions of chant and epic, tied as they were to mana-making hierarchies and status manifestation. Post-Enlightenment assumptions of constitutionalism and biblical morality on the part of her mentors—which she studied at sites such as the missionary-saturated Royal School, where she was enrolled as a youth, and at Kawaihao Church, where she would become choirmaster—were tinged with pathos. Foreign and Native mores came to complex integration in her poetry; it also informed a national project and neo-nativist mode of literary activism, which she carried on from her musically inclined cosmopolitan brother, whom Robert Louis Stevenson and other friends had called "the merry monarch" of the islands. His royal support of sugar plantations helped lead to his downfall.

Beyond demonstrating in the modernizing wake of Johann Gottfried von Herder, Ralph Waldo Emerson, and Henry Wadsworth Longfellow a "genius" for *Nationalliteratur*, poetry, narrative, and song became the deposed Queen Lili'uokalani's way to preserve the very ethos and ethnos of nationhood (Ka Lāhui [Hawaiian Nationhood]) founded in what she terms "patriotism, which for us [Hawaiians] means the love of the very soil on which our ancestors have lived and died" (38). What Queen Lili'uokalani calls her "Literary Occupation" during her imprisonment (chap. 54) affirmed the cultural-linguistic basis of Hawaiian Nationhood, as she outlines in chant and epic; such works also exposed the overextension of the US nation into the Pacific. As the *San Francisco Chronicle* noted of settler forces driving this "so-called revolution," which had earlier forced King Kalākaua to submit to a "Bayonet Constitution," the queen later sought to abrogate to her own undoing: "The government of the Sandwich Islands appears to have passed from the hands of the king into hands of a military oligarchy that is more domineering than Kalakaua ever was" (375).[21]

A "revolution" had been launched in 1893 "by the foreign element of the community," abrogating "the misrule of the native line of monarchs" and proclaiming "guarantees to the protection of life, liberty, and property," as the *New York Times* described the event in its issue dated January 28 of that year. Queen Lili'uokalani was deposed by self-proclaimed "revolutionary" forces of this new government under President Sanford Dole, which appealed, over the course of five years and in the wake of scathing censure by the Blount Commission's report during the administration of President Grover Cleveland, for US recognition in Washington. In her autobiography, the queen calls these white settler forces "pseudo-Hawaiians" and "so-called Hawaiians" (326) claiming to represent the Hawaiian nation in the remote eyes of Washington politicians. This pseudo-representation by settlers can still take place today. But the protests lodged in Honolulu and in the nation's capital (signed by some 38,000 of the 40,000 Hawaiian citizens of Native ancestry) still resonate. We need to hear these Pacific cosmopolitical voices seeking justice, as well as moral and political recognition, for the oceanic island land of Hawai'i.

By story and poem, as well as document and prayer, Lili'uokalani was writing *Hawaii's Story* not just to legitimate the statecraft of her reign, but also through appeals to international standards of justice, as well as the morality codes that American missionaries had brought into her country in 1820, to reclaim the nationhood of Hawai'i. Translating and writing poetry became her feats of ethnic transaction and Native policy. She recuperated the primordial Hawaiian heritage and language through the translation into English of the epic creation poem *Kumulipo*. "It is the chant which was sung to Captain Cook in one of the ancient temples of Hawaii," she noted, "and chronicles the creation of the world and of living creatures, from the shell-fish to the human race, according to Hawaiian traditions" (350). Her memoir, as well as her songs and poems (written in Hawaiian but with English and Spanish words at times deftly interspersed) constitutes a counter-record, a grievance, or a plea that resonates with place-specific allusiveness and tropological skills that befit a poet queen whose appeals to justice and moral idealism depended on American republicanism undone in contexts of imperial globalization.

Lili'uokalani's Hawaiian *mele* (song poem) "Ka Wiliwili Wai" (The Sprinkler) on the surface is a ditty that anthropomorphizes a lawn sprinkler near Washington Place: "Oh whirly-water / gentle rain shower on the move / what do you think you're up to / circling, twirling so quietly?" But the third stanza brings out the uncanny power of this modern convenience, disrupting ("taking over") harmonious relations among people, land, the heavens, and place: "Amazing / the way you take over: irresistible. / Come, slow down a little— / so I can drink!"[22] Another *mele* by Lili'uokalani, "Ka Wai Mapuna" (The Water Spring) conveys a more Hawaiian sense of space, where native waters flow calmly and "the Red Rain [after storms] brings peace."

The *Kumulipo*, published in 1889 by King Kalākaua under the title *He Pule Hoolaa Alii* (A Prayer Consecrating a Chief) and translated into English by Queen Lili'uokalani in 1897, was not just a long chant giving Hawaiian "aboriginal people" the status of a national literature with a worldview and creation epic; it worked politically to affirm the "supreme blood" claims of the Keawe-a-Heulu line as legitimate royal successors to the Kamehameha lineage (Kamehiro). It made this family direct descendants of the great fifteenth-century Chief Lonoikamakahiki and helped to quiet Hawaiian factions who refused to legitimate the claim to royal chiefly status of this family. King and queen were not just translating a poem of ethnographic interest to the Polynesian Society or the Bureau of American Ethnology (as the enterprising Nathaniel Emerson's *Unwritten Literature of Hawai'i: The Sacred Songs of the Hula* [1909] did). As brother-and-sister poets and indigenous myth makers, they were ratifying indigenous claims to nationhood by genealogical succession and cosmopoetic sanction as well as offering a capacious environmental vision of the indigenous ocean and land. As Kalākaua wrote in *Legends and Myths of Hawaii*, "Certain customs, like chants and meles, are matters of inheritance, and remain exclusively in the families with which they originate."[23]

Discussing how Hawaiian chants "were composed of symbolic phrases (*loina*) and hidden meanings (*kaona*)," the great nineteenth-century historian Samuel Kamakau surveys a catalogue of Hawaiian poetic and song genres (*mele*) that might fill Walt Whitman's "Chants Democratic" with wonderment facing their social range and efficacy:

> There were chants in honor of ancestors (mele kupuna), in praise of a land (mele 'aina), in praise of chiefs (mele ali'i), in praise of favorite children (mele hi'ilani), chants of gratitude (mele mahalo), chants of affection (mele aloha), chants of reviling (kuamu-amu), prayer chants (mele pule), dirges (kanikau), chants to put a person to sleep (mele hi-amoe), or to awaken one (mele ho'ala), chants asking a favor (mele noi), chants refusing the request (mele 'au'a), chants calling to be admitted (mele kahea), chants given as a gift (mele haawi), chants of boasting (mele ho'oki'eki'e), prophetic chants (mele wanana), chants foretelling events (mele kilokilo), chants of criticism (mele nemanema).[24]

The *Kumulipo* stands at the apex of this generic hierarchy of Hawaiian song and chant in its use of multilayered *kaona* (levels of meaning) and its coherent environmental cosmology linking the flora of sea and land together with Hawaiian presence.[25]

As the Hawaiian intellectual historian and politician David Malo portrayed in "He Inoa Ahi no Ka-la-kaua" (Fire Chant for King Ka-la-kaua), fire torches from Hilo to O'ahu had circled the island and illuminated the monarch's election in 1874 from the "Blessed fires of Mary of Peace" to

"Fire Company Number Two" to "the fire at [Iolani] Palace [that] shines in a circle, a full moon."[26] The poem performs a consecration of modern Honolulu city space by a sacred *kapu* (royal sanction). More than family heirloom or world epic, the *Kumulipo* was a genealogical possession, a generic and genetic power securely in the family possession, such as the right to burn *kukui* torches in the day. Aware of such genres and figurations of world-making import, Lili'uokalani compounded Christian and Native beliefs into what Helena Allen calls a synthesis of "Christian-Kahunaism."[27] Her moral path of nonviolent resistance and aloha was profound, as the kaona in this comment to her adopted daughter Lydia Aholo, which constitute a passage worthy of Saint Augustine, reveal: "To gain the kingdom of heaven is to hear what is not said, to see what cannot be seen, and to know the unknowable—that is Aloha. All things in this world are two; in heaven there is but One."[28] *Hawaii's Story* and poems of countermemory such as "The Sprinkler" can help us to hear the tale of Naboth's vineyard absorbed by the country of King Ahab that may have squandered the kingdom of heaven in its bargain for Pacific Empire. "Thus saith the LORD, In the place where dogs licked the blood of Naboth shall dogs lick thy blood, even thine" (I Kings 21:19), to invoke the biblical trope that gave Queen Lili'uokalani's written vision of a lost Hawaiian homeland the lasting force (*mana*) of allegory (kaona) as a literature of Oceania that will outlast us.

We need to rescue these internal forces of the Pacific from APEC-like formations of the future to make it serve different transcultural, pedagogical, historical, and "minor" uses tied to alternative energies of re-worlding. Allow me to end this chapter on such claims with a turn to the scholarship and poetry of Teresia Kieuea Teiawa, who did more to theorize and evoke the worlding of Oceania than anyone across the region, except for precursor writers such as Albert Wendt and Epeli Hau'ofa. Teiawa's works would move toward a different postmodern/postcolonial becoming in this important world region.

On Teresia Teaiwa's "Always Keeping It Real" in Oceania

We cultural workers and producers are part of a decades-long process of "keeping it real," to use the Pacific-based activating slogan of counter-discourse, countermemory, political-cultural contestation, and decolonizing energy that drives Teresia Teaiwa's poem "Amnesia," if not her whole writing-Oceania life as an engaged, located, theoretically adept poet-scholar. As Teaiwa affirmed in "Crisis Poem #2" a beautiful work from 2013 (as the crises were recurring, if not endless), "The world is a / poem that humans edit / ruthlessly."[29] We are engaged in a large-scale and situated making (worlding, reworlding) and unmaking (deworlding) of the British and the American Pacific as a kind of interlocking pragmatic imaginative and environmental set of imperial discourses and practices (both formal and informal) that took dominion, across world oceans, from the eighteenth-century Enlightenment era of contact (with its wars of light and dark) into the Romantic era of fascinated subsumption (circulated in erotic/exotic as well as missionary discourse)—discourses, practices, and mores that (to some extent) still hold sway (via settler-colonial and postcolonial formations) across our own era of amplified capitalist modernity, nuclearization, terrorism, and neoliberal globalization.

In her writing of essays and poems, Teaiwa was always *inventing concepts and tropes* to deworld (unmake) and reworld (remake) the given world, especially turning the Pacific Islands world into a planetary global Oceania (Gilles Deleuze: original philosophers and poets "invent concepts" / transfigurative tropes as a way to reframe and re-see the world).[30] As I have been trying not just to theorize but to materialize as a writing practice and a "becoming oceanic" poetics, we need a new world-island pedagogy that both learns and unlearns the Anglo-American canon of imperial possession and pragmatic expansionism; at the same time, the literatures of Pacific countermemory, de-creation, and reframing are implicated and surge into this pedagogical remaking.

Example A: "Militourism" putting together military infrastructure from World War II with global soft cultural tourism-*cum*-exotic primitivism of postwar global tourist culture. Example B: The "bikini" theorized and exposed by Teaiwa as a hypererotic trope of dominant Euro-American beach-blanket bodies, such as Brigitte Bardot and Marilyn Monroe, becoming seminaked and hypervisible at the same time that nuclear refugees of the Bikini Atoll atomic tests of 1945–1959 are unseen, occluded, rendered spectral or dead, and forgotten in the bikini. Example C: The literature and tactics of indigeneity, ethnonationalism, and racial tensions in the works of the Fijian writer Joseph Veramu and the Samoan writer Sia Figiel are considered by Teaiwa within translocal and transindigenous contexts of the "global war on terror."[31]

This mass-mediated war of Orientalizing abjection and ideological projection would seem to have little to do with the islander Pacific, but this war impinges (as she shows) via the collusions and outreach of contemporary US and British imperialism in an era of amplified *militourism*, (her term) as we see Fijian peacekeepers in the Golan Heights and US islander soldiers from American Samoa dying in Iraq and Afghanistan. "O beautiful! / O gracious skies! / O amber rays of Iraqi dust," as the Sia Figiel poem that Teaiwa quotes sardonically answers back. We see islanders spread out across global-local Pacific distances, grappling with the call to the Pax Americana.

Teaiwa's essay concludes with a vision of historical long-duration: "Fiji and American Samoa by virtue of their colonial pasts have both been drawn into the overlapping cartography of Middle Eastern conflicts, which have their roots in the history of the British Empire [but] their fruits in contemporary US imperialism." In all of her writing, both the far-reaching essays and the deft poems of satirical wit and lyrical beauty, at once oceanic and rationalized in form and trope, Teaiwa activates her tender modes of cultural-political pedagogy that both establish and destabilize the Anglo-American or Euro-Asia canon of imperial possession and pragmatic expansionism; at the same time, she helps to create and theorize the emergent literatures of Pacific countermemory, de-creation, and reframing as implicated in a multiscalar and thick-descriptive pedagogical remaking (or as I would affirm again) "reworlding" of ocean, people, language, and place. Here is one of her poems that enact this practice of projecting an Oceania that would challenge America's Asia Pacific as a hegemonic sublation of islands with another more archipelagic mode of belonging to the world of ocean and land:

> Searching for Nei Nim'anoa
> *September 24, 1994*
> *Suva, Fiji*
> I need to learn how to navigate,
> Read the stars, the wind, and the ocean swells
> Like she did.
> This drifting in a random sea of sites and sounds
> Has been too lonely.
> I will pick up the pieces
> of my broken Gilbertese,
> Gather the remnants of
> my broken heart,
> And use them to chart my course.

If I don't find her
Tao
She'll find me.[32]

A major framework such as "Pacific Rim modernism" often feeds on such minor expression and fears it like a strange pidgin speech that mocks and warps the standard currency. What minority groups construct inside of and against a major language such as English is not any essence but a movement, a position of unrest and agitation, a warping mode of "deterritorialization" that (at its best) outflanks and outwits the workings of the capitalist market, social regulation, and the nation-state as apparatus of canonical identity. Politicized by necessity (as was the indigenizing case with the Hawaiian queen's poetic and memoir writing and with the works of Teiawa), minority literature is tied to the collective enunciation, energy, and aspirations of a social grouping fed up with being excluded, dominated, or considered substandard, illiterate, or underdeveloped. Minority literature develops alternative forms, practices, and outlets that are the equal of the major languages; the major leagues of literature try to get into the minors, but they are often locked out and abandoned by minority forms.[33]

Displaced and nomadic at its core, that is to affirm with Deleuze and Chi'en, minority literature remains some mongrel mixture of *roots* and *wings* traced across huge multilingual spaces, such as the transpacific or Oceania. To invoke the mixed modes of aggravated diasporic affiliation in Jessica Hagedorn: "I'm not interested in just writing 'an American novel.' . . . Though I've been living in America for 30 years now, my roots remain elsewhere . . . , back there [in the Philippines]."[34] More so than British English today, "American English" has become a global language of minority becoming, a globalizing language of mongrelized complexity that goes on expanding the territory of global capital and the power of Americanization, but that is also emerging as a language that is being "deterritorialized" at multiple points and ties, which is to say, being worked over and transformed daily by minor expression and worked on by minorities and migratory forces via the interconnected worlding of internet connectivity.[35] American English is becoming Konglish and Singlish and Chinglish—that is to say, affirmatively—and is linked to Ebonics and Hawaiian Creole English. World English, unlike monolingual Microsoft-speak, is amplifying variations and mixtures at the world and local interface.[36] Minority literature may help to articulate these "borderlands" of open wounds, as well as the "borderwaters" of liquid interface across lands and oceans, forging languages of wry, tangled, bawdy, and blasted emergence where past and future mix and form something new, wild, strange: World English becoming Weird English.

At times I have pushed toward showing the routed rise of *diasporic* pulls across borders of nation and nation-state culture in the global era, as in sites such as San Francisco, Los Angeles, and Watsonville, California. I have otherwise advocated for forces revealing that a *rooting down* in culture and location takes place inside or against the nation, and more so across Oceania, as in the sovereignty struggle of Native Hawaiians, whose queen had been unjustly deposed by American settler forces. The pulls of the transnational and the local are quite strong now across Oceania, and we need to find different ways to articulate what national literature is in the process of becoming. Minor writing is not just what a minority writes. It is a distinctive kind of writing that aims to do something differently, in a language that is often experimental and risky in voice and angle of cultural-political vision. Milton Murayama does not aspire to become Jonathan Franzen of white middle-class anxiety and quest.

To invoke Deleuze and Félix Guattari on the social language experiment that is "becoming minor" in the full, decolonizing sense: "To make use of the polylingualism of one's own [major] language, to make a minor or intensive use of it, to oppose the oppressed quality of this language to its oppressive quality, to find points of non-culture or under-development, linguistic Third World zones by which a language can escape, an animal enters into things, an assemblage comes into play."[37] The Pacific Rim of oceanic interconnection is no longer below the radar or under the streets but is spreading across diverse geospatial sites in unexpected ways not even the idol managers could have planned for as capitalist spectacle of hybrid pop culture. *Asia becoming Oceania* is the figuration I would give to this pluralizing, as well as renativizing, dynamic where all is in the mix and flux of a transpacific coalition that Steve Martz tracks in *Ocean* (2020) as a "water world" of future possibility and what he calls the "wet globalization" of planetary immersion and eco-responsibility I have called *reworlding.*

Transplanted Poesis

Writing Oceania and the World

The ocean between California and Asia and the world oceans that make up much of this blue-green Earth credibly troped as Planet Ocean are the biomorphic and atmospheric elements that sustain and nourish us materially and aesthetically as a daily fact—indeed, as a moment-to-moment reckoning. At times, the Pacific in this study functions as an oceanic synecdoche of parts for whole, but at other times, oceans from the Indian and the Arctic to the Atlantic are touched on, as well as connected to, this indeterminate circumference of the ocean becoming Oceania. As Diane di Prima (1934–2020), the San Francisco–based Beat feminist poet, wrote in "Revolutionary Letter #2": "endless as the sea, not / separate, we die / a million times a day, we are / born a million times, each breath life and death."[1] In works from the radical broadsides of *Revolutionary Letters* (1971) through the dream texts of *Loba: Parts i–viii* (1978) to *The Poetry Deal* (2014), published by City Lights Foundation to honor her work as 2009 poet laureate of this coastal Silicon Valley city of queer, post-Beat, and techno-savvy progressiveness, di Prima embodied the oceanic sublime from the Atlantic of her birthplace to the Pacific of her alchemical poetics, radically enframed in a world metamorphosis through what she transmitted, in Buddhist terms, as *kind-king-mind*.[2]

As touched on in earlier chapters, Taiwan has become a crucial in-between site for me in this long-enduring, beatitude-questing self-formation (across US-linked continents and archipelagic world oceans) as a "becoming transnational"—or, more specifically, "transpacific"—scholar and poet. All is herein figured forth as post-Emerson and post-Kerouac "American scholars" in a worldly and worlding way to sites across Asia Pacific and Oceania; they could not have anticipated or transfigured into being in the eras of Manifest Destiny and the Cold War. Entangled not just with Mainland China and its emergent South Seas claims but also with the continental United States and its security-chain complex of financial buildup and military presence across the Pacific to East Asia, Taiwan challenges the present and future nation-state system as a complex in-between site of postmodern/postcolonial cultural poetics and its claims of an aggravated distinctive sort. Archipelagic and insular, Taiwan at once opens (and closes) to Mainland China and the mainland United States, as well as toward those stepping-stone islands (as Teresia Teaiwa mocked this instrumentalizing metaphor) across "Oceania." Taiwan is moving precariously toward figuring forth and materializing this dynamic of oceanic transformation as site/nation/locality/region in emergent ways of translocal solidarity and world belonging, as claimed in the field-reframing *Archipelagic American Studies* collection and Brian Russell Roberts's related *Borderwaters*.³

At the core of this vision of sympoiesis, world-literary enframing means learning an environmental blue/green ecopoetics articulated through thinking with and beyond the continental "mainland" claims and even the Asia imaginary of more nation-centered configurations or seeing Pacific Islands as separate states. This "becoming oceanic" poetics also suggests an attempt to overcome more absolutely racialized or rigidly territorially bordered frameworks of Asia and Pacific identity as spread across what Gerard Manley Hopkins, writing from "glass-blue days" on the global islands of England and Ireland, called the "the Wild air, world-mothering air / Nestling me everywhere."⁴

The obsessiveness of writing such worlding-poesis claims for a prefigurative "Oceania" that I have made in this study, and will continue to make and materialize in the years I have left on this planet, recalls the autoethnography embodied by Nathaniel Mackey, my UCSC colleague in poetry and poetics (now teaching at Duke University), who wrote of his multivolume *Double Trio*, "It was a period of distress and precarity inside and outside both. During this time, a certain disposition or dispensation

162

came upon me that I would characterize or sum up with the words all day music. It was a time in which I wanted never not to be thinking between poetry and music, poetry and the daily or the everyday, the everyday and the alter-everyday. Philosophically and technically, the work meant to be always pertaining to the relation of parts to one another and of parts to an evolving whole."[5] If the musical inspiration here shifted from Ornette Coleman's "A Love Supreme" and Miles Davis's "Flamenco Sketches" toward Bob Dylan's born-again collection *Trouble No More* and Handel's *Messiah* played over and over while rewriting this study, what Mackey calls the "alter-everyday" is another way of saying "reworlding." We both found inspiration in the "love's body" poetics and "metapolitics" of Norman O. Brown and the parts-whole pedagogy of Left Coast figures such as Robert Duncan and Jack Spicer, as is palpably explicit in the special issue of *boundary 2* we are both part of on Brown's future-projecting poetics of world collage and a kind of libidinal-gnostic metapolitics.[6]

As noted in chapter 3, worlding ecopoetics would reframe Hawai'i, Taiwan, Fiji, Okinawa, Cheju Island, and other sites as an affiliated poetics of transregional worlding, bioregion, and translocal solidarity enacted in "transpacific dharma wanderers" and writers such as Albert Saijo, Gary Snyder, Nanao Sakaki, Robert Sullivan, Susan Schultz, Juliana Spahr, and Craig Santos Perez. Poems I wrote in and about Taiwan refract some of this work through my personal trajectory as a National Science Research / Ministry of Science and Technology scholar and teacher in Taiwan from 1996 until the present in the global cities of Taipei, Hsinchu, and Kaohsiung, as linked to sites of research and cultural poetics at UCSC and the University of Hawai'i, Mānoa. Literature as invoked and theorized in a threefold process of *deworlding, worlding,* and *reworlding,* as discussed in interconnected sites along the Pacific Rim and across eco-emergent Oceania, can help us to grasp such environmental links among ocean, self, land, and planet. The Anthropocene demands as much if we are to flourish in these days of extreme weather and perilous transformations.

Transplanted Connect/I/Cut

"He who is transplanted sustains." That remains the motto of my home state of Connecticut in the northeastern New England region of the United States. it was posited to affirm an elected Euro-American community of believing elite, led by Thomas Hooker, who broke away from Massachusetts and transplanted to Wethersfield, Windsor, and Hartford in 1636 to found a new colony in the town Native Americans called Mattatuck (renamed Waterbury), where I was born. It was this latter-day moving sense of "*Connect/I/cut*," from the non-rooted to the *routed* outset, as Gilles Deleuze troped on this US state name, that later called out for on-the-road flight or perpetual breaking away from settled mores for newness, counter-conversion, and reinvention to happen.[7] This motto of diasporic affirmation and "cutting" across borders toward renewal could apply to the *cultural poesis* of outgoing and incoming writers to Hawai'i, as well, from the first bicultural Hawaiian convert, Henry Opukahai'a, and colonial mariner, John Ledyard, to the postmodern experimentalists Juliana Spahr and Susan Schultz, writing Hawai'i across discrepant eras. Being marked as a *haole* (white stranger) in this Native Hawaiian site, from the time of British, Russian, French, and American contact, became a colonial-settler problematic of not fully belonging to the language, place, history, bloodlines, or deeper culture of unsettled nationhood or statehood.

When I moved to Hawai'i in fall 1976, at the urging of the scholar-poet Josephine Miles, my English mentor at UC Berkeley, who had been a visiting creative writer (along with James Wright) during that summer, Hawaiian land struggles were ongoing, as were the battles to stop the bombing of Kahoolawe Island. Anti-tourist forces were mounting critiques in Kalama Valley and Waihole-Waikane Valley. The Hawaiian sovereignty struggles for land, ocean, language, and nation were in emergent states of articulation. In differently inflected if related terms, at the Talk Story conference and in the *Bamboo Ridge* journal (both in 1978), pro-localism and re-pidginized poetics clamored across the late 1970s. I was caught up in dialectics of this outsider-inside struggle, transformed, undone, remade, de-converted, immersed in crafty contradictions, baffled, amazed, and renewed as a writer and citizen of the world. Over the years of struggle for literary-political emergence, I came to call the English Department at the University of Hawai'i, Mānoa (in pidginized tones) "the Department of Anguish" and the beige building where I worked in office 513 "Ralph Kuykendall Hall," signaling how its namesake US liberal historian (like his literary coeditor, A. Grove Day) wrote magisterial histories and anthologies that absorbed Hawaiians into the telos of American statehood and all but cultural-political sublation into shopping mall citizens as James Michener's "Golden Men" of America Pacific hybridity.

While I came from the western part of Connecticut, the Nutmeg State of Yankee commerce, and grew up playing basketball and learning modes of working-class Catholicism and multiethnic tolerance in the Naugatuck Valley there as a half-Italian, half-Scottish byproduct, that was little help on the basketball courts of Kaniwae or when teaching in Mānoa. Soon entangled in the noisy blues of personal, transnational, and social transformation across those turbulent decades from 1976 to 2001, I had to learn my "Ananda Air Incorporated" cultural poetics—not to mention literary pedagogy, critical theory, and cultural studies—all over again. My work in poesis aimed to become a situated and retheorized mode of global, local, national, and regional belonging to Asia and the Pacific (*Asia Pacific*) as well as to Hawai'i (Native Hawai'i Nei) as a troubled homeland home. My ties to South Korea and Taiwan were prodded over the years through visits and residencies on the oceanic Rim, too.

The poems and books of transnational cultural studies reflecting late capitalist weather on the Pacific Rim I wrote and collections I coedited with scholars as diverse as Arif Dirlik, Wimal Dissanayake, Chris Connery, Soyoung Kim, Serena Chou, and Vilsoni Hereniko would reflect and refract these modes of transethnic, cosmopolitical, global-local, and transregional immersion. "Fear over the Ala Wai" is one poem that reflects that tension of being an outsider inside the dense commodity structures and ecological damage of tourism in Waikiki. "Seven Tourist Sonnets" aggravates the Euro-American sonnet form into modes of pidginized and mongrelized transnational-local belonging and displacement as settler-tourist poet on O'ahu island.[8] My students in beginning creative writing classes, such as "Da Pidgin Guerilla" Lee Tonouchi, the comedic Jennifer Waihe'e, and the soap actor and volleyball star Jason Olive, as well as many multicultural others over the years, gave me some of the languages, affects, and cultural mixtures that became fused into these poems.

The beneath-the-pavements community quested for, across these times and spaces as in a climate of poetic renewal and rebirth, stands for some "invisible republic" of becoming, *Tinfish*-like experimentation, multiple linkage, and minority transformation. "This is not exactly a people called upon to dominate the world," as Deleuze once affirmed of this nation written from below. "It is a minor people, eternally minor, taken up in becoming-revolutionary," at least for multitudes of immigrants, Beat-pilgrim souls, drifters, blues makers, border landers, protestors, border-water people of creative enunciation who would world the world into embodied ethos and link streets of the city to the oceans of the world, subterranean and submarine in some wondrous sense and sensibility we are on the verge of enunciating, transfiguring, materializing, and enacting as Oceania.[9]

The renewed mind as posited throughout this study, converted to a more poetic imagination of world belonging, can help transform the worldly world into the world where the metamorphosis of what I call "being always converting," always in quest for oceanic consciousness is open, is possible, can be made to happen as a reworlding process. If capital and its aligned forces go on deworlding the everyday world, we can link worlding to a reworlding dynamic as tied to dwelling more radically and organically as rooted in place, day by day. But also, reworlding means reflecting and refracting the becoming-multiple process of growth, regeneration that can activate new cells both biological and social. Such an oceanic poetics thus links us to the world-mothering air and world-mothering ocean to which figures as diverse as the poet-priest Gerard Manley Hopkins and environmentalist writers such as Rachel Carson, Rebecca Solnit, Diane di Prima, Teresia Teaiwa, and Gary Snyder have tried to awaken us via oceanic transfigurations of world belonging and world making across an array of urban sites and cultures. Oceanic becoming demands that we respect and learn from the ocean as a mode of ontological flourishing on this endangered and precious planet we share above and below the pavements of our far-flung cities and towns.

NOTES

INTRODUCTION. PACIFIC BENEATH THE PAVEMENTS

Epigraph: Clover, *Red Epic,* 3.

1 For an intertextually informed approach to the "blue humanities" as embodied phenomenology of oceanic swimming and belonging to this planetary element, see Mentz, *Ocean.* See also his study of oceanic antagonism
and commercial expansion in early modern contexts: Mentz, *Shipwreck
Modernity.*

2 See Mentz, *Ocean,* 17. He writes on Melville's indigenous Pacific hero
Queequeg as an oceanic being of advanced skills and empathy in the chapter titled "Melville's Aquaman" (95–102).

3 Weiss, "Driving the Beat Road."

4 See Phelan, "Italy's Plan to Save Venice from Sinking."

5 See McHugh, "Poet Ferlinghetti Chased Subs in World War II."

6 For a far-reaching study of the classical Mediterranean Ocean connecting Europe, Africa, and Asia, as later applied by Alexander von Hum-
boldt to enframe and trope the archipelagic Caribbean region as "seas
between lands" linked across the Americas, see Gillman, *American
Mediterraneans.* Gillman writes, "As such, this oceanic concept, as much
historical as ecological, offers a way to rethink comparative study three-
dimensionally, through the axis of space, time, and language" (xvi).

7 See R. Ghosh, "Albatross Unbound."

8   MacFarlane, "Marshland"; MacFarlane and Mitchell, "Hamburg's Spaces of Danger."

9   For the "heavy waters" of this nuclearized and militarized Pacific, as well as islands as spaces of environmental potentiality for "interspecies worlding," see DeLoughrey, *Allegories of the Anthropocene*, chaps. 2, 4. On geopolitics riddling the Indo-Pacific with the rise of Global China, see Hendrix, "America's Future Is at Sea."

10 See Roberts, *Borderwaters*, 1–43.

11 Somerville, "The Great Pacific Garbage Heap as Metaphor," 320.

12 Igler's *The Great Ocean* focuses on the eastern Pacific as a base for American-centered expansion. See also Huang, "Islands"; Shell, *Islandology*.

13 Hopkins, "The Blessed Virgin Compared to the Air We Breathe"; Wright, *Haiku*.

14 Probyn, *Eating the Ocean*, 2.

15 Probyn, *Eating the Ocean*, 38.

16 Haraway elaborates this cross-cutting process of "sympoiesis" via trans-species modes of "making-with" and creating yokes "for becoming-with" that proliferate multispecies "ongoingness" amid late-capitalist ruins: Haraway, *Staying with the Trouble*, 49, 58, 125. For desublimating formulations of "poison Apple" and its Silicon Valley nexus of labor to toxic hinterlands of China, see Litzinger, "The Labor Question in China."

17 For a summary of this discursive transformation occurring over the past two decades, see Suzuki, "Transpacific." See also C. Wang, *Transpacific Articulations*.

18 See Roberts and Stephens, *Archipelagic American Studies*, 1–4.

19 See Chan, "China Is Not Even Pretending Anymore in the South China Sea."

20 See *Economic Times*, "US Pacific Command Renamed as US Indo-Pacific Command." On the historical, cultural, and political formation challenges to this American Pacific, see Wilson, *Reimagining the American Pacific*.

21 *Economic Times*, "US Pacific Command Renamed as US Indo-Pacific Command."

22 Hsu, *A Floating Chinaman*, 12, 238 (the "floating" metaphor).

23 For a creative-destructive portrayal of the Shanghai Global telos at the vanguard edge of a global Chinese future and the Pearl River delta contado, see Connery, "Better City, Better Life."

24 Bernes, "Documents," 82.

25 See Pentland, "Vast Freshwater Reserves Discovered under Ocean Floor, Scientists Say."

26 Zanzibar B. Mcfate (@stonemirror) to the post-Adorno hypernihilist Eric Jarosinski (@NeinQuarterly), Twitter, posted October 19, 2014. The first version of this tweet, later translated into French, came from @NeinQuarterly in Paris from Berlin: "Under the paving stones: the beach. Under the beach: more paving stones. Under those paving stones: a street. Paved with beach."

27 See Pendell, *Walking with Nobby*.

28 On this Haus der Kulturen der Welt event in 2013, as linked to eco-aesthetic museum and industrial site interventions in Germany, Sweden, and Australia, see Robin, "Three Galleries of the Anthropocene."

29 See DeLoughrey, *Allegories of the Anthropocene*, chap. 4.

30 On the material history and allegorical trope of Tuvalu as disappearing world island, see DeLoughrey, *Allegories of the Anthropocene*, 178–90.

31 See Shanken, *Into the Void Pacific*. The catachresis of "the void Pacific" is taken from a letter of September 24, 1923, in which the British author D. H. Lawrence writes, "California is a queer place—in a way, it has turned its back on the world and looks into the void Pacific. It is absolutely selfish, very empty, but not false, and at least, not full of false effort": quoted in Shanken, *Into the Void Pacific*, 7. The World's Fair called for San Francisco to draw on Asian, the Pacific and Pacific Rim, and Latin American cultures (10, 25).

32 Wark, *The Beach beneath the Streets*.

33 Shepard and Smithsimon, *The Beach beneath the Streets*, 2.

34 Kim Stanley Robinson's science-fiction *Three Californias* and *Mars* trilogies depicting the "Pacific edge" of endangered life on the planet has given this working definition of *Anthropocene*: "The idea that we're living in the Anthropocene is correct. We are the biggest geological impact now; human beings are doing more to change the planet than any other force, from bedrock up to the top of the troposphere. Of course if you consider twenty million years and plate tectonics, we're never going to match that kind of movement. It's only in our own temporal scale that we look like lords of the Earth; when you consider a longer temporality, you suddenly realize we're more like ants on the back of an elephant" (Robinson, "California: The Planet of the Future"). See also Demos, "Welcome to the Anthropocene!"

35 In "The Politics of Seeing with the Global City," Nick Mirzoeff's description shifts from "world city" locality to the "global city": "In 2011, the world changed—statistically. For the first time, the majority of people worldwide lived in cities. That majority has already risen to 54% and is expected to reach 66% by 2050. What were previously known as 'world cities' during the era of high imperialism (1857–1945) depended on culture, trade, and war, following networks established by colonialism and directed by nation states. The global city is a metropolis whose primary relation is to other global cities, not to the nation where it is located. It is defined by its level of integration into the world-city network. From Shanghai to Johannesburg to New York, the global city has reconfigured itself 'back to the future.' It now looks more like 19th-century Paris than 1970s New York: a literally gilded center, epitomized by super tall buildings for the spectacularly wealthy, surrounded by extended zones of racialized hyper-segregation and poverty."

36 See bloomsby, response to "What specific event caused the Berlin Wall to fall peacefully?" Fun Trivia website, March 24, 2004, archived June 2,

2008, at Archive.org, https://web.archive.org/web/20080602160508/http://www.funtrivia.com/askft/Question45647.html.

37 For a portrait of the artistic and political ferment in anarchistic, urban-experimental, and Punk do-it-yourself (DIY) squats and sites such as the Kirche von Unten (Church from Below) in East Berlin and West Berlin before and after 1989, see Hockenos, "Zero Hour."

38 As quoted in Brislin, "On Berlin."

39 See Nichols, *Blue Mind*, 20.

40 On this movement in the humanities, see Gillis, "The Blue Humanities."

41 I allude to passages from Pendell, *Walking with Nobby*, in this study; see also Wilson, "Transfiguration as a World-Making Practice."

42 I touch more thickly on these movements in chapter 7 and the epilogue.

43 For an embrace of urban tactics of global capitalist disruption such as the Oakland Port blockage and Occupy "riot" tactics more broadly, from Watts and Bay Area streets to Paris and Egypt, see Clover, *Riot. Strike. Riot.*

44 Shepard and Smithsimon, *The Beach beneath the Streets*, 3.

45 See Benjamin, *Paris, Capital of the Nineteenth Century*, 157.

46 Gonick, "Global Cities and Their Discontents."

47 See Brown, "From Politics to Metapolitics"; Brown, *Love's Body*, 234 ("To liberate [urban] flesh and blood from reification, overthrow the [government by capitalist] reality-principle.").

48 See Miyoshi, *Trespasses*, 184.

49 Miyoshi, "Turn to the Planet," 286. On this environmental turn in Miyoshi's later writings, see Marran, "The Planetary."

50 Urbain, *At the Beach*, 118–19, 60.

51 See Teaiwa's frame-shattering essay on the nuclearized Pacific, "bikinis and other s/pacific n/oceans," and her posthumous collection, *Sweat and Salt Water*.

52 "Hong Kong's Taste for Seafood Putting Oceans in Danger."

53 Mentz, *At the Bottom of Shakespeare's Ocean*, 88.

54 The power of Jules Verne's transoceanic visions of spatial adventure, in French literary works of expanded globality such as *Around the World in Eighty Days* (1872) and *Twenty Thousand Leagues under the Sea* (1869), captivated Jean Cocteau's childhood imagination. Cocteau later observed, "Play and book alike [by Verne] not only thrilled our young imagination but, better than atlases and maps, whetted our appetite for adventure in far lands. . . . Never for me will any real ocean have the glamour of that sheet of green canvas, heaved on the backs of the Chatelet stage-hands crawling like caterpillars beneath it, while Phileas and Passepartout from the dismantled hull watch the lights of Liverpool twinkling in the distance": Cocteau, *Round the World Again in Eighty Days*, 1–2.

55 For thick-descriptive cultural-political studies of such phenomena and islander-based resistance, see Comer, *Surfer Girls in the New World Order*; Imada, *Aloha America*.

56 On the nuclear history and threat of rising ocean waters to the Marshall Islands, see the prescient ecopoetic collection, Jetnil-Kijiner, *Iep Jaltok*.

57 Lynas, *The God Species*, 5.

58 See Demick, "Malaysia Plane."

59 See Urbina, "Sea Slaves." As an example of "labor abuse at sea," Urbina portrays this transnational oceanic laborer in the South China Sea: "Mr. [Lang] Long's crews trawled primarily for forage fish, which are small and cheaply priced. Much of this catch comes from the waters off Thailand, where Mr. Long was held [by armed guards on a 'shoddy wooden ship'], and is sold to the United States, typically for canned cat and dog food or feed for poultry, pigs and farm-raised fish that Americans consume."

60 Marina Warner summarizes the hold of early modern cartographic images of oceans as monstrous sites in the fictions of world modernity, writing, "Moby-Dick, the giant squid and other sea monsters in Jules Verne, the Creature from the Black Lagoon, and even the computer-generated mask of glaucous, waving, suckered tentacles that the actor Bill Nighy is condemned to wear in the role of Davy Jones in *Pirates of the Caribbean: Dead Man's Chest* reveal the near-indestructible longevity of the marine world picture [of oceans as monstrous sites] that was crystallized by early cartographic visionaries": Warner, "Here Be Monsters."

61 Stein, *Jung's Map of the Soul*, 2, 9.

62 See Alaimo, "Violet-Black."

63 Whitman's poem can be read as one of the foundational American ecopoetic works: see Walt Whitman, "Out of the Cradle Endlessly Rocking," in Fisher-Wirth and Street, *The Ecopoetry Anthology*, 4–9. Whitman's will to American sublimity is connected to poem's coming to terms with the oceanic grandeur and waste of the Atlantic. On such postwar American ecopoetics, see the far-ranging arguments on poetic address and trope in Ronda, *Remainders*.

64 Fisher-Wirth and Street, *The Ecopoetry Anthology*, 49.

65 Mentz, *At the Bottom of Shakespeare's Ocean*, 18. Hereafter cited in the text.

66 See the essays in Cohen, *Prismatic Ecology*, esp. Tobias Menely and Margaret Ronda, "Red," 22–41.

67 Sloterdijk, *Neither Sun nor Death*, 239. Brown links "ut unim sint" (that they become) mandates to overcome division from Freud, Marx, and Christ to urge "our union with the sea (Thalassa); oceanic consciousness; the unity of the whole cosmos as one living creature": Brown, *Love's Body*, 82.

68 See Sloterdijk, *In the World Interior of Capital*, 5.

69 Alvin Lim, comment to Rob Wilson, "Rethinking World Literature" group site, Facebook, June 27, 2015.

70 Here I draw on Facebook input from the China-affiliated cultural studies scholars Ralph Litzinger and Meagan Morris, the latter a colleague on the editorial board of the journal *Inter-Asia Cultural Studies* since the mid-1990s.

71 Bill Knott, "On the Road," in Knott, *The Unsubscriber*, 42.

72 Pope Francis I, *Laudato Si*.

73 Santayana, *Persons and Places*, 59.

74 Lapham, "The Sea," 20–21.

75 Moore, *Plastic Ocean*, 24.

76 See Stamper, "At the Very Least Your Days of Eating Pacific Ocean Fish Are Over."

77 See the legalistic arguments in Schmitt, *The Nomos of the Earth in the International Law of the Jus Publicum Europaeum*.

78 Thorne, "The Sea Is Not a Place, or Putting the World Back into World Literature," 74.

79 See Wilson, "World Gone Wrong."

80 See DeLoughrey, *Allegories of the Anthropocene*; Schmitt, *Land and Sea*.

81 Urbain, *At the Beach*, 295.

82 Benjamin, *Reflections*, 98.

83 Schalansky, *Atlas of Remote Islands*, 8.

84 Schalansky, *Atlas of Remote Islands*, 74, 112.

85 Sloterdijk, *You Must Change Your Life*, 319.

86 Shell, *Islandology*, 188–90.

87 Spicer, *My Vocabulary Did This to Me*, 23.

88 Rankine, *Citizen*.

89 Ginsberg, *Collected Poems*, 880.

90 Ginsberg, *White Shroud*.

91 Among the semi-skeptical endorsements and crowd-sourcing tactics, see Kilvert, "Great Pacific Garbage Patch Plastic Removal System Could Become 'World's Biggest Piece of Marine Debris,' Critics Say," 6.

92 Somerville, "The Great Pacific Garbage Patch as Metaphor," 331.

93 Meltzer, "Interview with Kenneth Rexroth, Summer 1969," 30.

94 Eisen-Martin, *Heaven Is All Goodbyes*, 15. See also Eisen-Martin, *someone's already dead*.

95 For an interspecies portrayal as transfigured into bio-unity across Oceania, see DeLoughrey, "Ordinary Futures."

96 Abbas, *Hong Kong*. See also Estok and Chou, "The City and the Anthropocene."

CHAPTER 1. BECOMING OCEANIA

Earlier versions of this chapter were presented as talks at the Remapping Transnational American Studies panel, Modern Language Association Annual Convention, Boston, January 6, 2013; at the Sea Changes: Mediterranean and Maritime Perspectives on History and Culture conference, UCSC, May 3, 2013; the Social Life of Poetic Language *boundary 2* conference, University of California, Los Angeles (UCLA), April 12, 2014; and as the plenary talk at the Shifting Grounds: Cultural Tectonics along the Pacific Rim conference, Johannes Gutenberg University, Mainz-Germersheim, Germany, July 17, 2014. I have benefited from this feedback, as well as the research and book writing I was able to do as a Minister of Science and Technology visiting professor at Academia Sinica in Taipei, at the Institute of European and American Studies, which proved to be an exemplary site for "worlding Asia" and transpacific conversation.

1   For its focus on "wet globalization" and "blue environmentalism" after the works of Rachel Carson and Édouard Glissant, see Mentz, *Ocean*, 105–24. See also the blue-centered history of coastal belonging in Gillis, *The Human Shore*, and ocean-based social discourse studies in Steinberg, *The Social Construction of the Ocean*.

2   "Asia's Roiling Sea."

3   DeLoughrey, "Heavy Waters," challenges capital's "metaphors of liquid circulation."

4   Connery, "Pacific Rim Discourse," 36. Connery has reiterated claims that, "in both China and the United States, the oceanic imaginary has lost much of its force since 1989. In the United States, *Pacific Rim* is now mostly confined to fusion restaurants and universities": Connery, "Ideologies of Land and Sea," 196.

5   Melville, *Moby-Dick*, 468. On the longer duration of oceanic crossing as the perilous element of globalized capitalism and world exploration, moving from the globe-spanning Portuguese voyages of Vasco de Gama to the sinking of the *Pequod* in the Pacific and the capitalist-sublime *Titanic* in the North Atlantic, see Mentz, *Shipwreck Modernity*.

6   Olson, *Call Me Ishmael*, 114. Connery tracks this "as a journey that begins and ends in terrestriality to a pure trajectory, a pure oceanic deliverance": Connery, "The Oceanic Feeling and the Regional Imaginary," 303.

7   Reid, quoted in Cumings, *Dominion from Sea to Sea*, 140–41. For the "American Pacific" as a chain of military bases secured via coaling stations and harbor ports, see Eperjesi, "Basing the Pacific."

8   Kaplan, *Asia's Cauldron*, 13, 41. Dirlik has urged that we drop the Europe-originated use of "South China Sea" and "East China Sea" and instead use "SE Asia Sea" and "East Asia Sea," which were used historically by "dynastic regimes" in the region: Arif Dirlik, email to the author, May 27, 2014. On Chinese challenges to US hegemony and, to some extent, mimicry of some of its tactics and doctrines, see Mearsheimer, "The Gathering Storm."

9   See Schmitt, "The Monroe Doctrine as a Precedent for a *Grossraum* Principle." On the challenges to US oceanic supremacy by China, see also Hendrix, "America's Future Is at Sea."

10 On this world-altering Shanghai accord, see Kissinger, *On China*, 270.

11 See *New York Times*, "U.S., China Clash on Key Issues." As US Secretary of State Hillary Clinton avowed at the 2012 Pacific Forum, before she headed to tenser meetings with China in Beijing, "The Pacific is big enough for all of us": *Economist*, "Too Small an Ocean."

12 The Obama administration's ongoing "Pacific Pivot" is discussed in Copp, "Carter to Visit Beijing, Increase US-China Security Exercises."

13 See Stevenson, "China Makes Debut in RIMPAC Naval Drills."

14 Quoted in Helmreich, *Alien Ocean*, 3.

15 Nixon, "The Great Acceleration and the Great Divergence."

16 Spahr, *This Connection of Everyone with Lungs*.

17 Spahr, *Well Then There Now*, 84.

18 Spahr, *Well Then There Now*, 67.

19 McDougall, *The Salt Wind*, 3.

20 See McDougall, *Finding Meaning*, esp. chap. 3, wherein Native Hawaiian connections to earth, sky, and ocean are elaborated as ties to "ancestors" and a life full of pono (balance) and *mana* (spiritual life force).

21 Dimock, *Through Other Continents*, 4.

22 Santos Perez, *from Unincorporated Territory [hacha]*, 30; Santos Perez, *from Unincorporated Territory [saina]*, 11, 47.

23 Wilson, *Be Always Converting, Be Always Converted*, 87–142. On the plasticene ocean waste accumulation as linked to Asian sites and colonial aftermath, see R. Ghosh, "Albatross Unbound"; Liboiron, *Pollution Is Colonialism*, 72–73.

24 Meditating on the semiotics and politics of "Oceania," Hauʻofa admits having "Wansolwara" in mind, as the name of a newspaper produced by Pacific Islander journalism students at the University of South Pacific in Fiji when he founded the Oceania Centre for Arts and Culture in 1997, where the "Red Wave Collective" emerged: see Hauʻofa, *We Are the Ocean*, 114–17.

25 See the documentary portrait of Albert Wendt's oceanic trajectory from Samoa and New Zealand into the world of postcolonial literature in Shirley Horrock, dir., *The New Oceania*, Point of View Productions, 2005.

26 For the impact of Oceania as a framework and aspiration on diverse Pacific poets, see Wilson, "Postcolonial Pacific Poetries."

27 See Sandhu, "Allan Sekula."

28 Cumings, *Dominion from Sea to Sea*, 386.

29 See Connery, "Sea Power"; Connery, "There Was No More Sea."

30 Ben Child notes a PRC military officer's phobic allegorization of *Pacific Rim* as a military fable of oceanic domination, writing, "'The decisive battle against the monsters was deliberately set in the South China Sea adjacent to Hong Kong,' [PLA officer] Zhang wrote. 'The intention was to demonstrate the US commitment to maintaining stability in the Asia-Pacific area and saving mankind.' The Chinese military is currently concerned over US plans to transfer 60% of American naval assets to the Pacific by the

end of the decade. Zhang added: 'Soldiers should sharpen their eyes and enforce a "firewall" to avoid ideological erosion when watching American movies. More importantly, they should strengthen their combat capability to safeguard national security and interests'": Child, "Pacific Rim Designed to Advance US Cultural Domination of China."

31 See Gillis, "The Blue Humanities"; Mentz, *At the Bottom of Shakespeare's Ocean*, 97–99.

32 See Demick, "Malaysia Plane."

33 "Bathal, abyssal, and hadal zones [as] empty, void, null—an abyss of [deeper oceanic] concern," as Stacy Alaimo phrased "Violet Black," her entry on deep-ocean ecopoetics in Cohen, *Prismatic Ecology*, 233.

34 Sloterdijk, *Neither Sun nor Death*, 239. On the Germanic will to dominate the northern seas and islands from the era of *Hamlet* through the historical-rhetorical contexts of Napoleon, Hegel, and Hitler, and on the military-industrial, as well as cultural-political, Germanic drive to extend both southward to the Mediterranean and northward toward the seas of Denmark, Holland, and England, see Shell, *Islandology*, 183–248.

35 Sloterdijk, *In the World Interior of Capital*, 5.

36 Mentz, *Ocean*, 31–41.

37 Once defined by the range of a cannon shot from the shore, sovereignty over coastal waters has since 1982 been guided by the United Nations Convention on the Law of the Sea (UNCLOS). Signatories can claim a "territorial sea" up to 12 nautical miles (22 km) from their shoreline, inside which they can set laws but not meddle with international shipping.
   Beyond the territorial sea there is a 200-mile "exclusive economic zone" (EEZ), where coastal countries have the sole rights to resources. When two EEZs collide, UNCLOS calls for an equidistant line between the coasts, splitting the shared gulf or strait down the middle. The theory sounds simple, but the practice is complicated: islands, rocks, historic sovereignty and natural resources can bend the line. (*Economist*, "Make Law, Not War")

38 Kaplan, "Geography Strikes Back." On the genealogy of the South China Sea as both a key site of world trade and an emerging zone of political conflict, see Hayton, *The South China Sea*.

39 On the "neo-Mahanian" importance of "maritime geography" in contexts of a shrinking global commons and contemporary struggles over resource scarcity, see Hsiung, "Sea Power, Law of the Sea, and a Sino-Japanese East China Sea 'Resource War.'"

40 Helmreich, *Alien Ocean*, 31, 114–15.

41 This paragraph draws on information from Palmer, "Junk Accumulating on Monterey Bay Ocean Floors"; Shiels, "Boat Made of Trash [the Plastiki] Prepares to Set Sail."

42 See "John Luther Adams's 'Become Ocean' Wins Pulitzer Prize for Music,"
     WQXR blog, April 14, 2014: http://www.wqxr.org/#!/story/john-luther
     -adamss-embecome-oceanem-wins-pulitzer-prize-music.

43 George Yeo argues this contrast, writing, "The U.S. is, by self-
     identification, a missionary superpower. It judges others by its own
     standards and tries to shape them in the U.S.' own image—by hard and
     soft power. If China is also a missionary power, like the Soviet Union,
     perhaps a titanic struggle will again be inevitable. However, China is, by
     self-proclamation, not a missionary power. For China, a cardinal princi-
     ple of statecraft, not just the PRC but also its earlier incarnations, is
     non-interference in the internal affairs of others unless those affairs affect
     China's core interests": Yeo, "For the Rest of Asia, America Might Be a
     Friend, but China Cannot Be an Enemy."

44 Channer, "Spumante." Channer also writes and edits what he calls the
     genre of "Kingston noir."

45 For lyrics and a recording of this song, see Yasiin Bey (Mos Def), "New
     World Water," track 9 on *Black on Both Sides*, 1999, http://rapgenius.com
     /Yasiin-bey-new-world-water-lyrics.

46 Kerouac, *The Sea Is My Brother*, 11. Hereafter cited in the text.

47 Jack Spicer, "Second Letter," in Spicer, *My Vocabulary Did This to Me*,
     163.

48 Kerouac, *On the Road*, 408.

49 See Sacks, "In the Watery Part of the World."

50 Bob Dylan, "Lay Down Your Weary Tune," recorded October 24, 1963,
     Columbia Records, New York, released November 7, 1985, http://bobdylan
     .com/songs/lay-down-your-weary-tune.

51 Deleuze with Parnet, *Dialogues*, 51.

52 Barnett, "*The Sea Is My Brother*."

53 Barnett, "*The Sea Is My Brother*."

54 Carson, *The Sea around Us*.

55 This experimental world language poem activating the oceanic environ-
     ment (composed and signed "21 August 1960 Pacific Ocean at Big Sur
     California") is online at http://www.beatsupernovarasa.com/ARCHIVE
     /SEA.htm. Kerouac's *Big Sur* novel contains a long excerpt as an appendix:
     see Kerouac, *Big Sur*. Allen Ginsberg later wrote in a blurb for Penguin's
     1991 reprinting of Kerouac's alcohol-drenched novel, "Here at the peak
     of his suffering humorous genius, he wrote through his genius to end
     with 'Sea,' a brilliant poem appended, on the hallucinatory Sounds of the
     Pacific at Big Sur."

56 Kerouac's remark is quoted in David Henderson's introduction to
     Kaufman, *Cranial Guitar*, 15.

57 See Roberts, *Borderwaters*. See also the oceanic deformation of national
     borders and island linkages in Roberts and Stephens, *Archipelagic Ameri-
     can Studies*.

1 Dimock, *Weak Planet*.

2 When Tani Barlow, editor of the journal *Positions* and longtime Asia scholar, saw Soyoung Kim's flyers announcing the Worlding Asia, 1 and 2, conferences at Korea National University of the Arts in mid-November 2020, she responded to my circulation of this call with a rejoinder, "Why not Asianizing the world [rather than 'worlding Asia']?" Her prodding has led to this distinction here. As for Euro-theory-derived models of "worlding" from Hegel, Marx, Goethe, Arendt, and Derrida, this is the top-down theory approach applied in Cheah, *What Is a World?* This chapter contends that "worlding Asia" is set against its Euro-theory model, which gets applied by Cheah to "postcolonial novels" from the Global South.

3 For a challenge to the Anthropocene as an androcentric and heteronormative framework, from postnuclear and transhuman grounds of queer microbial forms of relationality and invertebrate temporalities and forms, see Kirksey, "Queer Love, Gender Bending Bacteria, and Life after the Anthropocene." See also the reframing of Anthropocene into Caribbean "Plantationocene" in DeLoughrey, *Allegories of the Anthropocene*, 33–62.

4 See Alfaisal, "Epistemic Reading and the Worlding of Postcolonialism."

5 See, e.g., Cheah, *What Is a World?*; Hayot, *On Literary Worlds*.

6 See Wilson and Connery, *The Worlding Project*.

7 Nancy, *The Creation of the World or Globalization*, 34, 117 fn. 2. Reflecting on Marx's critique of capitalist-driven globalization, Nancy defines the world as "a totality of meaning" (41), especially for "those who inhabit it" (42). To create the world actively as a struggle is Nancy's aim: "*To create the world* means: immediately, without delay, reopening each possible struggle for a [just] world" (54).

8 See Haraway, "A Giant Bumptious Litter."

9 In the words of the French translator Betsy Wing, Glissant's "world is totality (concrete and quantifiable [*la totalité-monde*]), echoes (feedback [*les échos-monde*]), and chaos (spiraling and redundant trajectories [*le chaos-monde*]), all at once, depending on our many ways of sensing and addressing it": Glissant, *Poetics of Relation*, xv. Glissant urges the proliferation of worlding echoes as against chaotic "globality": "*Échos-monde* thus allow us to sense and cite the cultures of peoples in the turbulent confluence whose globality organizes our chaos-monde" (94). As Dash argues about this Earth-ocean poetics, "Archipelago has become Glissant's quintessential concept of dwelling in the world": Dash, "The Stranger by the Shore," 356–57.

10 Glissant urges a contrast between "rooted identity" as becoming intolerantly fixed and "relation identity" as becoming multiple, more ecologically connected to other relations on the planet: Glissant, *Poetics of Relation*, 142, 146. On Glissant's approach to intrarelationality and proliferating

relation as "what the world makes and expresses of itself," see Kaiser, "Worlding Comp Lit." By tactile analogy to Glissant's mode of worlding, Kaiser invokes Barad's formulation, "The world theorizes as well as experiments with itself. Figuring, reconfiguring": Barad, "On Touching—The Inhuman That Therefore I Am," 207.

11 This is what Cheah tracks in his theory-*cum*-novel-driven study *What Is a World?* as the "world-impoverishing and world-alienating" effect of capitalist globalization, as challenged by "worlding" forms of postcolonial novels": Cheah, *What Is a World?*, 96. For post-Heideggerian tactics of worlding within modernity, see Hayot, *On Literary Worlds*.

12 See Ong, "Introduction," 11. For *worlding* as "a word, an argument, and a possibility" to a range of digitally impacted sites and forms from video-game communities to internet shopping to world traveling, see Trend, *Worlding*, viii–x.

13 Nancy, *The Creation of the World or Globalization*, 109.

14 See Lowrie, "Why the Phrase 'Late Capitalism' Is Suddenly Everywhere."

15 Harvey, *Seventeen Contradictions and the End of Capitalism*, 246. Hereafter cited in the text.

16 Stevens, *Opus Posthumous*, 249.

17 For a far-ranging infrastructural, as well as ideological, critique of this Chinese formation, see Kuo, "One Belt, One Road." See also sublimated world domination as portrayed in Chin, *Savage Exchange*. On *Snowpiercer* as figuration of the capitalist-driven Anthropocene, see Rob Wilson, "*Snowpiercer* as Anthropoetics."

18 J. Kim, *Postcolonial Grief*, 5. What Kim terms the "dread forwarding" of war trauma is compounded by the ecological dread forwarding of the Anthropocene effects in the Pacific.

19 Wark, *Molecular Red*, xii.

20 On this Trump discourse during his first Asia visit, see Wilkinson et al., "Trump's New Foreign Policy Touts a Free and Open Indo-Pacific." On the Jeremaic genre of geopolitical transfiguration and critique, see Wilson, *Be Always Converting, Be Always Converted*, 7, 9, 18–19, chap. 5.

21 A. Ghosh, *The Great Derangement*, 92.

22 Environmental disasters may just create further market opportunities for another mode of "disaster capitalism"-*cum*-enforced "-austerity" to proliferate, from Bangladesh and Beijing to West Virginia and back: see Harvey, *Seventeen Contradictions and the End of Capitalism*, 249. Naomi Klein documents and contests such views in *This Changes Everything*, 106–10.

23 June Wang, "Worlding through *Shanzhai*: The Evolving Art Cluster of Dafen in Shenzhen, China," in Wang et al., *Making Cultural Cities in Asia*, 125–27.

24 Nancy, *The Creation of the World or Globalization*, 37, 117.

25 Tsing, *The Mushroom at the End of the World*, 20.

26 Haraway, *Staying with the Trouble*, 49, 58, 125.

27 See the cognitive mapping of this distended ecoplanetary sublime in Morton, *Hyperobjects*.

28 Wilson and Connery, *The Worlding Project*, 216. On the phenomenological origins of *worlding* in Heidegger and postcolonial transformations of this approach in Spivak and Said, as well as post-9/11 geopolitical formations of "world literature" canons, see D'haen, "Worlding World Literature."

29 Hayot, *On Literary Worlds*, 6.

30 Tsing, *The Mushroom at the End of the World*, 21. For an inter-Asian approach to this "transpacific" dynamic of exchange, see Yao, "Oceanic Etymologies."

31 See Gluck and Tsing, *Words in Motion*, 15.

32 Tsing, *The Mushroom at the End of the World*, 139.

33 Cho, "Remembering Lucky Dragon, Re-membering Bikini," 127.

34 Lepawsky, *Reassembling Rubbish*, 1–48, 129–49.

35 Lepawsky, *Reassembling Rubbish*, 5–6.

36 In "Making Kin," the biology metaphor-laden talk Donna Haraway gave at the Center for Cultural Studies on the University of California, Santa Cruz, campus on January 31, 2004.

37 Hayot, *On Literary Worlds*, 31.

38 Quoted in Associated Press, "China's Largesse in Tonga Threatens Future of Pacific Nation."

39 Trend, *Worlding*, 5.

40 Trend, *Worlding*, 170.

41 Wilson and Connery, *The Worlding Project*, 213, 216.

42 Barad, "Quantum Entanglements and Hauntological Relations of Indifference," 265, emphasis added.

43 Barad, "On Touching—The Inhuman That Therefore I Am," 207.

44 C. Wang, "Dreams of Colliding Worlds," 21.

45 Hayot, *On Literary Worlds*, 39. See the overview of this globalizing-worlding dynamic as a European center-periphery contradiction in Juvan, "Worlding Literatures between Dialogue and Hegemony."

46 See Radhakrishnan, *History, the Human, and the World Between*.

47 Solnit, *Wanderlust*, 29.

48 Wallace Stevens, "Of the Surface of Things," in Stevens, *The Collected Poems of Wallace Stevens*, 57.

49 *Worlding* remains underspecified and untranslated, close perhaps to Nancy's preferred French term *mondialisation* as meaning "world-forming," as if echoing *welten*, that Germanically freighted noun-become-verb from Heidegger's existential ontology: Nancy, *The Creation of the World or Globalization*, 117.

50 Alon, "The Becoming Literature of the World."

51 Alon, "The Becoming Literature of the World."

52 See Cheah, "What Is a World?"

53 Cheah, *What Is a World?*, 192.

54 Cheah, "World against Globe," 319.

55 Cheah, *What Is a World?*, 330.

56 Cheah, *What Is a World?*, 331.

57 Alon, "The Becoming Literature of the World."

58 Ong, "Introduction," 11.

59 Ong, "Introduction," 13.

60 Roy and Ong, *Worlding Cities*, xv; Ong, "Introduction," 11.

61 Chen, *Asia as Method*, xv, 5.

62 Ong, "Introduction," 4.

63 Goldman, "Speculating on the Next World City," 236–37, emphasis added.

64 Z. Lee, "Eco-cities as an Assemblage of *Worlding* Practices." (Lee draws his concept of Asian-based "worlding" from Ong.)

65 Z. Lee, "Eco-cities as an Assemblage of *Worlding* Practices," 186.

66 Siu, "Retuning a Provincialized Middle Class in Asia's Urban Postmodern," 131.

67 Chua, "Singapore as Model," 48.

68 Haines, "Cracks in the Façade," 171.

69 Ong, "Introduction," 11.

70 Ong, "Introduction," 13.

71 Chen, *Asia as Method*.

72 Roy, "Conclusion," 309.

73 Roy, "Conclusion," 314.

74 Roy, "Conclusion," 327.

75 Roy, "Conclusion," 331.

76 Goh, *Worlding Multiculturalisms*, 8.

77 Goh, *Worlding Multiculturalisms*, 3.

78 Goh, *Worlding Multiculturalisms*, 5

79 Chong, "Bukit Brown Municipal Cemetery," 178.

80 Ho, "Casino Multiculturalism and the Reinvention of Heritage in Macao," 135; S. Kim, "Shopping Mall as Dwelling-Place," 158.

81 See Wilson and Dirlik, *Asia/Pacific as Space of Cultural Production*.

82 See Hauʻofa, *We Are the Ocean*, 56.

83 Hauʻofa, *We Are the Ocean*, 50–51.

84 On Hauʻofa's "counter-conversion" from a Pacific developmental telos to an oceanic polytheism and planetary belonging, see Wilson, *Be Always Converting, Be Always Converted*, 119–42.

85 See Jeff Orlowski, dir., *Chasing Coral*, Netflix Documentary, 2017; Joseph Orlowski, dir., *Chasing Ice*, Exposure Production, National Geographic, 2012.

86 "Scientists are unequivocal about the cause of the [coral] bleaching: Our oceans are warming, because they are absorbing more and more greenhouse gases as humans release massive, harmful amounts of carbon into the atmosphere. The more they heat up, the more algae the coral polyps must release to ensure their own short-term survival—but coral can't survive long-term in such warm temperatures. It's estimated that close

to half of the coral in the Great Barrier Reef alone have died in the last 18 months": Lapin, "'Chasing Coral.'"

87 Since 2000, sea temperatures around Korea have risen 1.3 degrees Celsius, resulting in a huge decrease in, and loss of, marine creatures, including dietary species such as snow crab, squid, mullet, and abalone, to name just a few in this cuisine-rich context in Seoul: see Chyung and Park, "Rich Marine Species Vanish with Climate Change."

88 See Rice, "Coral Reefs at Severe Risk as World's Oceans Become More Acidic."

89 See Rice, "Coral Reefs."

90 Purdy, *After Nature*, 267.

CHAPTER 3. TOWARD A BLUE ECOPOETICS

Epigraph: Spillers, "Mama's Baby, Papa's Maybe," 72. As Tiffany Lethabo King summarizes this transnational "black oceanic" as variously material-ized and troped, "Hard to escape, the ocean and its legacy has crested again and again in Hortense Spillers's notion of the 'oceanic,' Édouard Glissant's 'archipelagic thought,' Paul Gilroy's 'Black Atlantic,' Kamau Brathwaite's 'tidalectics,' Antonio Benítez-Rojo's 'rhythm,' and Omise'eke Natasha Tinsley's 'Black Atlantic, queer Atlantic'": King, *The Black Shoals*, 5.

1 For an overview of this formation in poetry, see Wilson, "Postcolonial Pacific Poetries."

2 Blum et al., "Introduction," 5.

3 As the editors of this special issue of the *Journal of Transnational American Studies* argue, "Such modes of analysis understand the planet as contingent and kinetic, and thus archipelagic and oceanic orientations may have special explanatory power in an age of ecological crisis": Blum et al., "Introduction," 9.

4 Arac, "Aesthetics in the Discipline of Literary Studies Today."

5 The trope of "transpacific dharma wanderers" (reflecting shared interests in ecology, Buddhism, and the environment, as discussed later) alludes to the title of a poetry reading given by Gary Snyder, Nanao Sakaki, and Albert Saijo at the University of Hawai'i, Mānoa, on March 2, 2000, organized by Richard Hamasaki and his colleagues: see Burlingame, "Wanderers Find Poetry in the Environment."

6 See Brathwaite, *Middle Passages*; Philip, *Zong!*; Spillers, "Mama's Baby, Papa's Maybe."

7 Along these ecopoetic lines, see the essays in DeLoughrey and Handley, *Postcolonial Ecologies*.

8 Adelbert von Chamisso, *Remarks of a Naturalist*, as quoted in Igler, *The Great Ocean*, 3, 129–42. Chamisso's voyage into the Pacific of 1815–18 registered, as Igler writes, how "naturalists witnessed native groups not at

a moment of early contact, but instead two or more generations after the epidemics, community fragmentation, and instruments of colonialism had severely altered the lives and physical conditions of native populations" (132)—hence, an earlier phase of Pacific "deworlding."

9 See Wilson, "Reframing Global/Local Poetics in the Post-imperial Pacific"; Wilson, *Reimagining the American Pacific*.

10 See Walker, *The Country in the City*, esp. chap. 5.

11 This paragraph draws on Palmer, "Junk Accumulating on Monterey Bay Ocean Floors"; Shiels, "Boat Made of Trash [the Plastiki] Prepares to Set Sail"; Simons, "Weather Eye."

12 See Mentz, *An Introduction to the Blue Humanities*.

13 On "Hauʻofa's Hope" as vision of the re-indigenizing Pacific "necessarily entangled with other, more ambivalent, scenarios and dystopias," see Clifford, *Returns*, 212.

14 "The image, the imagined, the imaginary—these are all terms which direct us to something critical and new in global cultural processes: *the imagination as social practice*. No longer mere fantasy (opium for the masses whose real work is elsewhere), no longer simple escape (from a world defined principally by more concrete purposes and structures), no longer elite pastime (thus not relevant to the lives of ordinary people) and no longer mere contemplation (irrelevant for new forms of desire and subjectivity), the imagination has become an organized field of social practices, a form of work (both in the sense of labor and of culturally organized practice) and a form of negotiation between sites of agency ('individuals') and globally defined fields of possibility": Appadurai, "Disjuncture and Difference in the Global Capitalist System," 5.

15 Said, *Orientalism*, 6.

16 In "Conversion," Pierre Hadot elaborates classical and post-Christian meanings of conversion from *epistrophe* (return of self to an origin) to the radical version of *metanoia* (mutation and rebirth).

17 Wilson, *Be Always Converting, Be Always Converted*.

18 Wilson, *Be Always Converting, Be Always Converted*, 126.

19 Wilson, *Be Always Converting, Be Always Converted*, 126.

20 Brown, *Love's Body*, 81.

21 From Saint Augustine to Ralph Waldo Emerson and Marshall McLuhan (as in his televised talk to Father Peyton), God was commonly defined as "a circle whose center is everywhere and whose circumference is nowhere."

22 Pompallier, *Early History of the Catholic Church in Oceania*.

23 As Félix Guattari elaborates on the Greek origin of the term, *ecology* derives from a complex of "house, domestic property, habitat, natural milieu," meaning the geo-place where interactions and encounters take place: Guattari, *The Three Ecologies*, 91n52.

24 Snyder, *Earth House Hold*, 127.

25 On a "God-term" (such as money) as designating the ultimate motivation or substance of a given rhetorical or constitutional frame, see Burke, *A Grammar of Motives*, 355.

26 Hauʻofa, *We Are the Ocean*, 30–36.

27 "For a day or two we have been plowing [in the Pacific near Fiji] among an invisible vast wilderness of islands, catching now and then a shadowy glimpse of a member of it. There does seem to be a prodigious lot of islands this year; the map of this region is freckled and fly-specked all over with them": Twain, *Following the Equator*, 72.

28 See Winduo, "Chief of Oceania," in a cluster of memorial works by colleagues and friends titled "Epeli's Quest: Essays in Honor of Epeli Hauʻofa."

29 See Hu-DeHart, *Across the Pacific*; Lowe, *The Intimacies of Four Continents*; Saldivar, *Trans-Americanicity*; Shu and Pease, *American Studies as Transnational Practice*.

30 On modes of regional transformation and cultural studies, see Lyons, "American Pacific Culture and Theory"; Wood, "Cultural Studies for Oceania."

31 Wilson and Dirlik, *Asia/Pacific as Space of Cultural Production*, 6.

32 For related interventions into Pacific refiguration, see Dirlik, *What Is in a Rim?*; Hereniko and Wilson, *Inside Out*. For a techno-cyberspace take, see Hjorth and Chan, *Gaming Culture and Place in Asia-Pacific*.

33 Wilson and Dirlik, *Asia/Pacific as Space of Cultural Production*, 13.

34 Chen, "Taiwan as Club 51: On the Culture of US Imperialism," 111.

35 See Dirlik, "Asia Pacific Studies in an Age of Global Modernity." Hereafter cited in the text.

36 Meltzer, "Interview with Kenneth Rexroth, Summer 1969."

37 Hauʻofa, *We Are the Ocean*, 118.

38 For historical and aesthetic contexts for such experimental-local work, see Wilson, "Pacific Postmodern."

39 Hauʻofa, *Tales of the Tikongs*. Hereafter cited in the text.

40 Hauʻofa, *Kisses in the Nederends*.

41 Connery, "Pacific Rim Discourse," 40.

42 Hauʻofa, *We Are the Ocean*, 56.

43 Hauʻofa, *We Are the Ocean*, 50–51.

44 Hauʻofa, *We Are the Ocean*, 53.

45 Vilsoni Hereniko, ally in Pacific cultural studies and director of the Pacific Studies Center at the University of Hawaiʻi and successor to Hauʻofa as director of the Oceanian Arts Center at the University of the South Pacific in Suva, offered feedback on the issue as to how, and to what extent, Hauʻofa "excluded" Asians from Oceania or the material national history of the Pacific. Vilsoni pointed out, at my keynote talk to the twenty-first Annual School of Pacific and Asian Studies conference in 2010, through historical anecdote and critical reflection, that Hauʻofa did *more than anyone* to

support the Indians in Fiji, scapegoated Indo-Fijian settlers at a time when other Pacific writers were more on the side of keeping Fiji for Fijians and supporting the nativist-based hegemony and regime changes there. Thus, too, less capacious forms of Pacific and Oceania-based "identity" had often become a way of including/excluding "Asia" (another vague term) and Asians (like those Indo-Fijians of four and five generations labor in Fiji) from linking the "subaltern regional Pacific" around art culture and social vision and ecology, etc.

46 Teresia Kieua Teaiwa, "Amnesia," in Teaiwa, *Sweat and Salt Water*, 6. Teaiwa performed "AmneSIA" with the Samoan novelist Sia Figiel for the compact disc *Tereneisa* (Eleipaio, Honolulu, 2000).

47 Pule, *The Shark That Ate the Sun*, 73–75. For ongoing struggles against this militarized archipelago of US bases across the Pacific, see Shigematsu and Camacho, *Militarized Currents*.

48 See Yamashita, *Anime Wong*; Yamashita, *I Hotel*.

49 Inada, "Shrinking the Pacific."

50 Balaz, "Polynesian Hong Kong."

51 "In globalization, there are no cultures, but only nostalgic images of national cultures: in postmodernity we cannot appeal back to the fetish of national culture and cultural authenticity. Our object of study is rather Disneyfication, the production of simulacra of national cultures; and tourism, the industry that organizes the consumption of those simulacra and those spectacles or images": Jameson, "New Literary History after the End of the New," 379.

52 Wilson, *Be Always Converting, Be Always Converted*, 139.

53 For indigenous-affiliated transpacific linkages and alternative ecological frameworks toward people, place, and nation, see P. Huang, *Linda Hogan and Contemporary Taiwanese Writers*.

54 See H. Huang, "Performing Archipelagic Identities in Bill Reid, Robert Sullivan and Shyman Raporgan," 281.

55 H. Huang, "Towards Transnational Native American Literary Studies."

56 Snyder, *Earth House Hold*, 140–42. See also Sakaki, *Break the Mirror*; Sakaki, *How to Live on Planet Earth*.

57 Snyder, *Earth House Hold*, 143.

58 Saijo, OUTSPEAKS, 73–74, 163. Hereafter cited in the text.

59 See Santos Perez, *from Unincorporated Territory [hacha]*; Santos Perez, *from Unincorporated Territory [saina]*; Santos Perez, *from Unincorporated Territory [guma]*.

60 Santos Perez, *from Unincorporated Territory [hacha]*, 10.

61 Santos Perez, *from Unincorporated Territory [hacha]*, 47.

62 Santos Perez, *from Unincorporated Territory [saina]*, 44.

63 Santos Perez, *from Unincorporated Territory [saina]*, 63.

64 Santos Perez, *from Unincorporated Territory [saina]*, 113.

65 Santos Perez, *from Unincorporated Territory [saina]*, 113, For a book-length poem re-creating Maori Oceania, see Sullivan, *Star Waka*.

66 Santos Perez, *from Unincorporated Territory [saina]*, 47.

67 Santos Perez, *from Unincorporated Territory [saina]*, 87.

68 Santos Perez, *from Unincorporated Territory [saina]*, 115.

69 Santos Perez, *from Unincorporated Territory [saina]*, 119. See also Teaiwa, 1994.

70 If globalization presumes that "world space" at the mercy of market norms and neoliberal policies is reshaping the world from Beijing to Paris, this can lead to what Jean-Luc Nancy calls earth-shattering values of the *immonde* or "glomus" delivered to the planet by this world-becoming-market: Nancy, *The Creation of the World or Globalization*, 37, 117.

71 Wilson, *Be Always Converting, Be Always Converted*, 15.

72 Spivak, *Death of a Discipline*, 85.

73 Snyder, *A Place in Space*, 233.

74 Snyder, *A Place in Space*, 255.

75 Snyder, "Coming into the Watershed," 234.

76 Snyder, *The Old Ways*, 3–6.

77 Snyder, *A Place in Space*, 6.

78 Clifford, "Indigenous Articulations," 22–23.

79 Clifford, "Indigenous Articulations," 22–23.

80 Clifford, "Indigenous Articulations," 31.

81 See Wendt et al., *Whetu Moana*, 6–10, 122–25.

82 Buell, *Writing for an Endangered World*, 22. For oceanic and riverine ecopoetics, see also chap. 6 ("Global Commons as Resource and as Icon: Imagining Oceans and Whales") and chap. 8 ("Watershed Aesthetics").

83 Wilson, *Be Always Converting, Be Always Converted*, 139.

84 I. Frazer (from Dunedin, New Zealand) captures this implicit sense of coalition building in *We Are the Ocean* in a customer review written for Amazon.com on March 19, 2009: "[This collection] will be of interest to everyone who knows his work and who shares his optimism and passion for the Pacific—which he preferred to call, in the spirit of *pan-Pacific cooperation and inclusiveness*, Oceania" (emphasis added).

85 Quoted in Hau'ofa, *We Are the Ocean*, 52.

86 Quoted in Wilson, "Milton Murayama's Working Class Diaspora across the Japanese/Hawaiian Pacific."

87 For Asian-affiliated dimensions of Richard Hamasaki's work as poet, see Wilson, "Review of Richard Hamasaki's *From the Spider Bone Diaries: Poems and Songs*."

88 Wilson and Dirlik, *Asia/Pacific as Space of Cultural Production*, 345–49. For a related approach to Reuney's seafaring poetics, see Peter, "Chuukese Travellers and the Idea of Horizon."

89 Wilson and Dirlik, *Asia/Pacific as Space of Cultural Production*, 347.

90 Chuukese is an Austronesian language of the Malayo-Polynesian family branch dispersed across the Pacific islands, Indonesia, Malaysia, and the Philippines. Thus far it is more dispersive than the related yet

Taiwan-concentrated branch of the Austronesian Formosan languages used by pre-Han peoples of Taiwan.

See Alan E. Davis's Chuuk lexicon at Davis, "A Preliminary List of Animal Names in the Chuuk District, with Notes on Plant Names."

91 See, e.g., Peter, "Chuukese Travellers and the Idea of Horizon."

92 Wilson and Dirlik, *Asia/Pacific as Space of Cultural Production*, 349.

CHAPTER 4. MIGRANT BLOCKAGES, GLOBAL FLOWS

Epigraph: Mackey, *Splay Anthem*, x.

1 Nail, *The Figure of the Migrant*, 1.

2 Jameson's figurations of the "postmodern sublime" in relation to capitalist-situated formulations and aesthetic-political projects are discussed in Wilson, "The Postmodern Sublime."

3 See Wellman and Landau, "South Africa's Tough Lessons on Migrant Policy."

4 See Pai, *Scattered Sands*.

5 See Reddy and Vimalssery, *The Sun Never Sets*.

6 Nixon, *Slow Violence and the Environmentalism of the Poor*, 19, 152.

7 Žižek, "Have Michael Hardt and Antonio Negri Rewritten the *Communist Manifesto* for the Twenty-First Century?," 229.

8 For a fuller discussion of these and other Dylan songs, in transnational and national contexts, see Wilson, "Bob Dylan in China, America in Bob Dylan."

9 Bourne, "Trans-national America."

10 Marx, *The Communist Manifesto*, 49.

11 Žižek, "Have Michael Hardt and Antonio Negri Rewritten the *Communist Manifesto* for the Twenty-first Century?," 227.

12 Sassen, *Guests and Aliens*, 2.

13 Bauman, *Globalization*, 87. See also Sassen, *Expulsions*.

14 Harvey, *Seventeen Contradictions and the End of Capitalism*, ix–x.

15 Harvey, *Seventeen Contradictions and the End of Capitalism*, 156.

16 Harvey, *Seventeen Contradictions and the End of Capitalism*, 174.

17 See Urbina, "Sea Slaves."

18 As quoted in Okeowo, "The Writing Life of a Young, Prolific Poet."

19 On differing immigration regimes in Germany, France, and Italy as they evolved in relation to seasonal, as well as capitalist and intra-state, shifts in situation and policy, see Sassen, *Guests and Aliens*, chap. 4,

20 Sassen, *Guests and Aliens*, xiii.

21 Sassen, *Guests and Aliens*, xvi.

22 Sassen, *Guests and Aliens*, xx, 151.

23 See Barkan and Shelton, *Borders, Exiles, Diasporas*.

24 One result of this global flow impact has come to be called "the increasingly dominant definition of the postcolonial as *migrant*," one result being

the widespread writing of a polyglossic cosmopolitan literature in the metropolis genres of London, New York, and Toronto: see Boehmer, *Colonial and Postcolonial Literature*.

25 Sassen suggests that this figure of "denizenship" emerged in Western European research from 1990 as "a way of giving immigrants the full range of rights without the necessity to acquire a new citizenship": Sassen, *Guests and Aliens*, 146.

26 On this "call to globality" and embrace of "the creativity of desire" in the multitudes, see Hardt and Negri, "Marx's Mole Is Dead!"

27 Greater China global patterns are traced in Tally and Lee, *Remapping the Homeland*. These flows were aggravated by the global pandemic crisis of 2020–21 that originated in Wuhan but spread through discrepant medical and social policies from Italy to the United States.

28 Clifford, *Routes*, 249.

29 Sassen, *Guests and Aliens*, 136.

30 Nail, *The Figure of the Migrant*, 235.

31 Sassen, *Guests and Aliens*, xv, 150–57. Sassen urges that, despite the varying causes and patterns, "there is a geopolitics of migration" (150).

32 Ruthlessly simplifying the problem of global migrancy and its causes and consequences, Australia has a zero-tolerance policy toward migrants who arrive by sea, allegedly to prevent drownings such as those now occurring on Mediterranean shores.

33 As Nail argues, "A continually increasing population of migrants, with partial or no status, who are subject to a permanent structural inequality (the lack of voting and labor rights, possible deportation, and other deprivations depending on the degree of status) is difficult to reconcile with almost any political theory of equality, universality, or liberty": Nail, "Alain Badiou and the *Sans-Papiers*."

34 See Foucault, "The Refugee Problem Is a Presage of the Great Migrations of the Twenty-first Century."

35 See Nguyen, *The Displaced*; Nguyen, *Nothing Ever Dies*; Nguyen, *The Refugees*; Nguyen, *The Sympathizer*. Nguyen also coedited one of the first collections to examine "transpacific" critically as an emergent transoceanic framework: see Nguyen and Hoskins, *Transpacific Studies*.

36 Gilroy, *After Empire*, 165.

37 Gilroy, *After Empire*, 166–67.

38 Gilroy, *Postcolonial Melancholia*, 149.

39 On formations of social care in the Philippines as affected by US migration and citizenship policies toward Filipino Americans, see Choy, *Empire of Care*.

40 See Andersson, "The European Union's Migrant 'Emergency' Is Entirely of Its Own Making."

41 A. Kim, "Satellite Images Can Harm the Poorest Citizens."

42 Nail, *The Figure of the Migrant*, 3.

43 Arendt, "We Refugees," 117.

44 See Mardorossian, "From Literature of Exile to Migrant Literature."

45 Leslie, "Walter Benjamin."

46 Roy, "Edward Snowden Meets Arundhati Roy and John Cusack."

47 Aldo Leopold, "Game and Wildlife Conversation" (1930), as quoted in Nixon, *Slow Violence and the Environmentalism of the Poor*, 240.

48 See Walters, "Author Junot Diaz Called Unpatriotic as Dominican Republic Strips Him of Award."

49 Anzaldúa, *Borderlands / La Frontera*, 3.

50 Donald Trump demeaned all such immigrants as "criminals and rapists" in fall 2015; hence, legislative attempts to stop sanctuary city ordinances have been called "the Donald Trump Act" to discredit them. On coming to terms with the haunted and traumatic legacies of the Japanese American internment and its far-flung urban and transpacific diaspora, see Karen Tei Yamashita's uncanny experimental text on her San Francisco and Los Angeles maternal and paternal families and the academic disciplines ("letters") that would grasp such displacements: Yamashita, *Letters to Memory*.

51 E. Lee, *The Making of Asian America*, 9.

52 E. Lee, *The Making of Asian America*, 13.

53 Kingston, *Tripmaster Monkey*, 4. Hereafter cited in the text.

54 Kerouac, *The Portable Jack Kerouac*, 233.

55 Kingston, *The Fifth Book of Peace*, 66.

56 Snyder, *A Place in Space*, 3–5. Snyder's original essay on North Beach as multicultural "habitat" was published in 1977; Yamashita's "Ai Hotel" chapter of *I Hotel* satirically comes to terms with so-called white dharma bums of North Beach and Chinatown (such as Snyder and Kerouac) who cultivated white Buddhism-*cum*-Tantric yoga as a mode of subjective revolution.

57 Yamashita, *I Hotel*, 2. Hereafter cited in the text.

58 On these Taiwan-PRC tensions over territory and identity that get acted out in the makings of Asian American studies in the United States during this period, see C. Wang, *Transpacific Articulations*.

59 C. Wang, *Transpacific Articulations*, 9.

60 E. Lee, *The Making of Asian America*, 283.

61 Nail, "Alain Badiou and the *Sans-Papiers*," 123.

62 Haraway, "Anthropocene, Capitalocene, Plantationocene, Chthulucene / Making Kin," 160. The politics of neoliberal care and faith-based compassion have interacted to inform the humanitarian rise of "church asylum" as sacred space of refuge for immigrants in contemporary Europe: see Mitchell, "Freedom, Faith, and Humanitarian Governance."

Second epigraph: "Diane Di Prima's 'Rant.'" Di Prima read the poem, which is about world making as soul making, as part of her acceptance speech for the Fred Cody Lifetime Achievement Award in April 2012.

1   Urban greening requires the alteration of urban energy systems and this planetary turn: see Castan Broto and Robin, "Climate Urbanism as Critical Urban Theory."

2   Dissanayake and Wilson, *Global/Local,* calls attention to this place-altering transnational dynamic that still happens and requires a shift in the "field imaginary" of disciplines.

3   See Leonard, "Our Plastic Pollution Crisis Is Too Big for Recycling to Fix." Kenya is now leading the way in a far-reaching ban of plastic use: see Watts, "Eight Months On, Is the World's Most Drastic Plastic Ban Working?"

4   Brechin, *Imperial San Francisco.*

5   On Jobs's counterculture fascination with Dylan, as documented by his biographer Walter Isaacson, see Greene, "New Steve Jobs Bio Reveals Details of His Relationship with Bob Dylan and Bono."

6   See Wilson, "Spectral City."

7   Spahr and Buuck, *An Army of Lovers,* 13. On Big Science and infrastructural knowledge production of the atomic bomb and atomic energy at UC Berkeley, see Brechin, *Imperial San Francisco,* chap. 5.

8   Kaufman, *Solitudes Crowded with Loneliness,* 87.

9   On these nuclear displacements and cancerous damage, see the "History Project" section in Jetnil-Kijiner, *Iep Jaltok,* 13–31.

10 For a study of this nuclearized American Pacific as centered in the archipelagic islands of Hawai'i, see Wilson, "Postmodern as Post-nuclear."

11 Ginsberg, *Howl and Other Poems,* 39.

12 Spahr and Buuck, *An Army of Lovers,* 101.

13 Kaufman, *Solitudes Crowded with Loneliness,* 11.

14 Taylor, *Eldorado,* 240–41.

15 Meltzer, "Interview with Kenneth Rexroth, Summer 1969."

16 Paul Kantner, "On the Jefferson Airplane in San Francisco," *San Francisco Chronicle,* January 29, 2016, https://libquotes.com/paul-kantner/quote/lbz7b4n. Kantner was the guitarist for the rock bands Jefferson Airplane and Jefferson Starship.

17 As quoted in Greene, "New Steve Jobs Bio Reveals Details of His Relationship with Bob Dylan and Bono."

18 Greene, "New Steve Jobs Bio Reveals Details of His Relationship with Bob Dylan and Bono."

19 Brechin, *Imperial San Francisco,* xxiii.

20 Brechin, *Imperial San Francisco,* 27.

21 Walker, *The Country in the City,* xii–xii, 6.

22 Brechin, *Imperial San Francisco*, 249.

23 See the photographs in Solnit, *Hollow City*. See also the reworlding urban atlas maps in Solnit's *Infinite City*, assigned repeatedly in my San Francisco courses at UCSC.

24 Brechin, *Imperial San Francisco*, xxii.

25 See Snyder, *The Practice of the Wild*.

26 Snyder, *A Place in Space*, 233.

27 Snyder, *A Place in Space*, 234, 255.

28 Snyder, *The Old Ways*, 3–6.

29 Snyder, *The Old Ways*, 3.

30 Snyder, *The Old Ways*, 5–6.

31 See Trend, *Worlding*.

32 See Kerouac, *Desolation Angels*.

33 Kerouac, *The Dharma Bums*, 105.

34 Kerouac, *The Dharma Bums*, 99.

35 Kingston, *The Fifth Book of Peace*, 66.

36 Buell, *Writing for an Endangered World*, 22.

CHAPTER 6. HIROSHIMA SUBLIME

Epigraphs: Vidal, "Requiem for the American Empire," 15; Ōe, "The Myth of My Own Village," 144.

1 See "The Nuclear Doomsday Clock."

2 For the nuclear trauma interpreted as a mutual American/Japanese national fantasy of domination and victimization, see Pease and Wilson, "A Conversation with Ōe Kenzaburō." For nuclear trauma and the wars in Vietnam and the Persian Gulf, see Pease, "Hiroshima, the Vietnam Veterans War Memorial and the Gulf War."

3 See Wilson, *American Sublime*, 228–63.

4 Wolfe connects this Japanese fascination with the atomic bomb and a suicidal will to self-sublation into Empire, death, and national rebirth, as well as a countermovement, within emerging narratives of postmodernity to activate "a specifically Japanese nuclear criticism" no longer fascinated with the techno-euphoria of Western modernity: see A. Wolfe, *Suicidal Narrative in Modern Japan*, 223. See also Ferguson, "The Nuclear Sublime." On the American will to dominate nature and space as sublime, see Pease, "Sublime Politics."

5 Ōe, *Hiroshima Notes*, 19. Describing the global culture of postmodernity, Andreas Huyssen invokes the need for a social-psychological process of *Vergangenheitsbewaltigung* (coming to terms with the past) to show how collective mourning was belatedly activated in postwar Germany by works such as the popular American TV series *Holocaust*. Overcoming "a

whole network of mechanisms which aimed at repressing, denying, and making unreal (*Entwirklichung*) Germany's Nazi past," a process of social mourning took place in which a release of the repressed was aroused and Germans were able, finally, to identify with the Jewish victims of their will to nation-state superiority. I do not mean to argue historical identity between the postmodern American and German situations, by any means, but I do want to suggest that, through "reimagining Hiroshima," a release of the American repressed could take place, and we could better negotiate the trauma's victims lest we stayed locked within the techno-euphoria of war: see Huyssen, *After the Great Divide*, 94–114.

6 Ballard, *Empire of the Sun*, 366. Hereafter cited in the text.

7 Žižek, *The Sublime Object of Ideology*, 203. On nuclear fantasies still operative in the militarization of the Pacific and the Gulf War, see Wilson, "Postmodern as Post-nuclear"; Wilson, "Sublime Patriot." Residual associations of the sublime with icons of technology are surveyed in Nye, *American Technological Sublime*, esp. chap. 9. A rocket launch for Nye captures the American sublimation of technology into an icon of national power, especially after Hiroshima had traumatized the national subject: "Here millions crowd together seeking an experience that can powerfully represent national greatness" (254).

8 Hillman, "Wars, Arms, Rams, Mars."

9 Morris, "Politics Now (Anxieties of a Petit-bourgeois Intellectual)," 19. On the "fantasy-organization of desire" within the postmodern political imaginary of the nation-state sublime, see Žižek, "Eastern Europe's Republics of Gilead"; Žižek, *Looking Awry*, 162–69.

10 Žižek, *Looking Awry*, 165.

11 Ibuse, "The Crazy Iris."

12 Quoted in Weart, *Nuclear Fear*, 103.

13 See Sakai, "Return to the West / Return to the East."

14 Prior to the imperial outreach into Asia that led to Japanese confrontation with the United States, Karatani links Japanese discovery of the sublime as national aesthetic possibility to the internal colonization of Hokkaido by the Japanese nation-state. "For the vast wilderness of Hokkaido inspired awe in human beings, unlike the mainland which had been regulated for centuries and enveloped by literary texts": Karatani, *Origins of Modern Japanese Literature*, 41.

15 Ōe, *Hiroshima Notes*, 78. Hereafter cited in the text.

16 Ōe, *A Personal Matter*. Other fiction by Ōe concerned with the direct (or displaced) trauma of nuclear horror includes *The Silent Cry* (1974) and the anthology *The Crazy Iris and Other Stories of the Atomic Aftermath* (1985).

17 For the nuclear trauma as mutual American/Japanese national fantasy of domination and victimization, see Pease and Wilson, "A Conversation with Ōe Kenzaburō."

18 Duras, *Hiroshima Mon Amour*.

19 Perlman, *Imaginal Memory and the Place of Hiroshima*, 106.

20 Fussell, *Thank God for the Atom Bomb and Other Essays*, 25.

21 Fussell, *Thank God for the Atom Bomb and Other Essays*, 24.

22 Fussell, *Thank God for the Atom Bomb and Other Essays*, 19–20, 24.

23 Fussell, *Thank God for the Atom Bomb and Other Essays*, 29.

24 Fussell, *Thank God for the Atom Bomb and Other Essays*, 17, 25.

25 Fussell, *Thank God for the Atom Bomb and Other Essays*, 34.

26 Fussell, *Thank God for the Atom Bomb and Other Essays*, 26.

27 Ibuse, *Black Rain*, 59–60.

28 Ibuse, *Black Rain*, 190–91.

29 Hull, "'Hiroshima: Out of the Ashes' Takes Point of View of Victims."

30 Quoted in Ōe, *Hiroshima Notes*, 155.

CHAPTER 7. WAKING TO GLOBAL CAPITALISM AND OCEANIC DECENTERING

1 See Ch'ien, *Weird English*; Jameson, *A Singular Modernity*. On these trans-pacific interactions of literary modernity spreading across the region from Paris to Tokyo and San Francisco and back, see Park, *Apparitions of Asia*.

2 Wilson, *Pacific Postmodern*. Ch'ien's *Weird English* takes a comparable approach to "Chinglish" as a literary language of social becoming.

3 For a strong example of such a locally based yet Hawaiian valued poetics, see Balaz, *Pidgin Eye*.

4 In 2004, APEC held its twelfth summit, this time in Santiago, Chile, the first time for a meeting to be held in South America since APEC's formation in 1989. One of its organizing founders, along with Australia, Korea attaches high importance to APEC's neoliberalist concept of "open regionalism," especially since trade with APEC's twenty members accounted for some 68 percent of Korea's total trade in 2004. See Garikipati, "APEC Provides Platform for Free Trade, Investment." For APEC, the "Asia-Pacific region" very much means what we and most other global shapers and agents would call the "Pacific Rim."

5 This stance of pidgin voice and Hawaiian ethos used by Balaz in his poetry and editing is tracked in Wilson, *Reimagining the American Pacific*.

6 See Mackey, *Discrepant Engagement*, 27.

7 On the shifts of San Francisco configured as a Pacific Rim city of post-Beat culture and techno-driven newness, see Wilson, "Spectral City."

8 This is what Asada Akira traced, in its Japanese postmodern version, as "infantile capitalism," based on the perpetual neo-differentiation of niches, linkages, desires, and codes: Asada, "Infantile Capitalism and Japan's Postmodernism."

9 Irving, "Rip Van Winkle," 1349.

10 Jean Baudrillard, *America*, quoted in Iwabuchi, *Recentering Globalization*, 41.

11 Connery, "Pacific Rim Discourse," 50–56.

12 The promise of capitalism's creative-destructive ethos drives the market-geared urban forms and shopping mall cities of China, as outlined in the Harvard Design School Project on the City: see Chuihua et al., *Project on the City I*.

13 On rural-urban imbalances in globalizing India, see Spivak, *Other Asias*, chap. 5. See also the globalization tactics and national crisis management of postsocialist Chinese knowledge/power as mapped in H. Wang, *China's New Order*, esp. sec. 3.

14 For a "translational" version of "transnational literacy" and "crisis management" in US area studies, ethnic studies, and comparative literature, see Spivak, *Death of a Discipline*, esp. chap. 3.

15 On US Orientalism, see Edward Said's defamiliarization of postwar liberal humanities and social sciences: Said, *Orientalism*, 284–328. He tracks the discourse of US State Department–driven "American Orientalism" as the formation of area studies in the 1950s. See also Colleen Lye's archival work in *America's Asia* on the racial, state, and class origins of area studies in Jack London, Frank Norris, and others, as well as the self-Orientalizing formation in David Henry Hwang's *M. Butterfly*.

16 Dirlik, *What Is in a Rim?*, 15–36.

17 Woo-Cumings, "Market Dependency in U.S.-East Asian Relations." See also Johnson, *Blowback*; Johnson, *The Sorrows of Empire*.

18 See Coffman, *Nation Within*, 289–313.

19 Lili'uokalani, *Hawaii's Story by Hawaii's Queen*, 316–17. Hereafter cited in the text.

20 See Silva, *Aloha Betrayed*, 190.

21 The text is excerpted in Lili'uokalani, *Hawaii's Story by Hawaii's Queen*, app. A.

22 Translated from the Hawaiian by Mary Kawena Pukui, in Pukui and Korn, *The Echo of Our Song*, 171–72.

23 Lili'uokalani, *Hawaii's Story by Hawaii's Queen*, 334–49.

24 Kamakau, *Ruling Chiefs*, 240.

25 On this Native Hawaiian poetics of wordplay and world-making vision, see McDougall, *Finding Meaning*, esp. chap. 2.

26 Pukui and Korn, *The Echo of Our Song*, 142–43.

27 Allen, *Betrayal of Lili'uokalani*, 19

28 Allen, *Betrayal of Lili'uokalani*, 401–2.

29 See Teaiwa, *Sweat and Salt Water*.

30 Teaiwa (1968–2017) was born in Honolulu to an I-Kiribati father and an African American mother and was raised in Fiji. She was the author of the poetry collection *Searching for Nei Nim'anoa* (1995) and coauthor, with

Vilsoni Hereniko, of *Last Virgin in Paradise: A One-Act Play* (1993). Her creative work was also published as the compact disc *Terenesia: Amplified Poetry and Songs* (2000). Teaiwa earned a Bachelor of Arts from Trinity College in Washington, DC; a master's in history from the University of Hawai'i; and a doctorate in the history of consciousness from the University of California, Santa Cruz, where she formed the core of a Native Pacific Edge community of scholarship. She taught history and politics for five years at the University of the South Pacific in Suva, Fiji, before moving to New Zealand to teach Pacific studies at Victoria University. She died in 2017, to the great sorrow of many across the Pacific, from Suva and Honolulu to Santa Cruz and Kaohsiung.

31 Teaiwa, "Reading Imperialism in the Pacific."

32 Teaiwa, *Searching for Nei Nim'anoa.*

33 "Minority" and "minor" are not the same thing, but the minority can remain tied to the minor literature and minor genres and see this not as a failure but as a distinctive achievement.

34 Hagedorn, 'The Exile Within / The Question of Identity."

35 "The first and most fateful deterritorialization is then this one, in which what Deleuze and Guattari call the axiomatic of capitalism decodes the terms of the older precapitalist coding systems and 'liberates' them for new and more functional combinations. . . . [T]here comes a moment in which the logic of capitalism—faced with the saturation of local and even foreign markets—determines an abandonment of that kind of specific production, along with its factories and trained workforce, and, leaving them behind in ruins, takes its flight to other more profitable ventures [such as new forms of financial speculation shorn from land and place]. . . . Globalization is rather a kind of cyberspace in which money capital has reached its ultimate dematerialization": Jameson, *The Cultural Turn*, 152–54.

36 On innovative and trans-normative Chinglish forms of "Chinky Writing," see Ch'ien, *Weird English*, chap. 2. On the deterritorializing of "American English" being deployed "in the service of minorities," Third World linkages, and counter-national becomings that defy normative canonization or social control, see Deleuze with Parnet, *Dialogues*, 58.

37 See Deleuze and Guattari, *Kafka*, 27. In a review essay written for a special issue of *Postcolonial Studies* (2008), coedited by Christopher Leigh Connery and Venita Seth, I survey the novelistic trilogy of Milton Murayama as portraying a rooted and routed labor diaspora of Japanese Hawaiian Americans moving across the twentieth century from Japan to Hawai'i and to California and New York City and back: Wilson, "Milton Murayama's Working Class Diaspora across the Japanese/Hawaiian Pacific." In other words, Murayama's novels offer a worlded, interacting, and transoceanic Pacific Rim vision.

1   Di Prima, "Revolutionary Letter #2," quoted in Stein, "Diane di Prima's *Revolutionary Letters*," 8.

2   On this and her radically oceanic vision of consciousness and community, see Stein, "Diane di Prima's *Revolutionary Letters*."

3   See Roberts and Stephens, *Archipelagic American Studies*, esp. 34–35, 281–302.

4   See Hopkins, "The Blessed Virgin Compared to the Air We Breathe."

5   Mackey, quoted in "Double Trio: Poetry by Nathaniel Mackey," *New Direction*, n.d., https://www.ndbooks.com/book/double-trio/#.

6   See Mackey, "Song of the Andoumboulou"; Wilson, "Transfiguration as a World-Making Practice."

7   "'Connecticut, Connect-I-cut!' cries little Joey" in Deleuze and Guattari, *Anti-Oedipus*, 37.

8   See Wilson, "Seven Tourist Sonnets."

9   Deleuze, *Essays Critical and Clinical*, 4. See also the Pacific-situated eco-poetic essays in Huang and Lin, *Pacific Literatures as World Literature*.

BIBLIOGRAPHY

Abbas, Ackbar. *Hong Kong: Culture and the Politics of Disappearance*. Minneapolis: University of Minnesota Press, 1997.

Alaimo, Stacy. "Violet-Black." In *Prismatic Ecology: Ecotheory beyond Green*, edited by Jeffrey Jerome Cohen, 233–51. Minneapolis: University of Minnesota Press, 2013.

Alfaisal, Haifa Saud. "Epistemic Reading and the Worlding of Postcolonialism." *Transmodernity* 7, no. 2 (2017): 41–56.

Allen, Helena G. *Betrayal of Liliʻuokalani*. Honolulu: Mutual, 1991.

Alon, Shir. "The Becoming Literature of the World: Pheng Cheah's Case for World Literature." *Los Angeles Review of Books*, December 19, 2016. https://lareviewofbooks.org/article/becoming-literature-world-pheng-cheahs-case-world-literature.

Andersson, Reuben. "The European Union's Migrant 'Emergency' Is Entirely of Its Own Making." *Guardian*, August 22, 2015. http://www.theguardian.com/commentisfree/2015/aug/23/politics-migrants-europe-asylum.

Anzaldúa, Gloria. *Borderlands / La Frontera: The New Mestiza*. 5th ed. San Francisco: Aunt Lute, 1999.

Appadurai, Arjun. "Disjuncture and Difference in the Global Capitalist System." *Public Culture* 2 (1990): 1–24.

Arac, Jonathan. "Aesthetics in the Discipline of Literary Studies Today." *Politics/Letters*, July 18, 2020. http://politicsslashletters.org/commentary/aesthetics-in-the-discipline-of-literary-study-today.

Arendt, Hannah. "We Refugees." In *Altogether Elsewhere: Writers on Exile*, edited by Marc Robinson, 111–19. Boston: Faber and Faber, 1994.

Asada, Akira. "Infantile Capitalism and Japan's Postmodernism." *South Atlantic Quarterly* 87 (1988): 629–34.

"Asia's Roiling Sea." *New York Times*. August 18, 2012. http://www.nytimes.com /2012/08/19/opinion/sunday/asias-roiling-sea.html.

Associated Press. "China's Largesse in Tonga Threatens Future of Pacific Nation." *Star Advertiser* (Honolulu), July 8, 2019. https://www.staradvertiser.com /2019/07/09/breaking-news/chinas-largesse-in-tonga-threatens-future-of -pacific-nation.

Balaz, Joe. *Pidgin Eye*. Kaneohe, HI: Ala, 2019.

Balaz, Joe. "Polynesian Hong Kong." *Otoliths* 16 (February 1, 2010). http://the -otolith.blogspot.com/2010/01/joe-balaz-polynesian-hong-kong-its.html.

Ballard, J. G. *Empire of the Sun*. New York: Pocket, 1984.

Barad, Karen. "On Touching—The Inhuman That Therefore I Am." *Differences* 23 (2012): 206–23.

Barad, Karen. "Quantum Entanglements and Hauntological Relations of Indif- ference: Dis/Continuities, SpaceTime Enfoldings, and Justice-to-Come." *Derrida Today* 3 (2010): 240–68.

Barkan, Blazar, and Marie-Denise Shelton, eds. *Borders, Exiles, Diasporas*. Stan- ford, CA: Stanford University Press, 1998.

Barnett, David. "*The Sea Is My Brother: The Lost Novel*, by Jack Kerouac" (review). *Independent*, November 27, 2011. http://www.independent.co.uk/arts -entertainment/books/reviews/the-sea-is-my-brother-the-lost-novel-by -jack-kerouac-6268430.html.

Bauman, Zygmunt. *Globalization: The Human Consequences*. New York: Colum- bia University Press, 1998.

Benjamin, Walter. *Paris, Capital of the Nineteenth Century*. In *Reflections: Essays, Aphorisms, Autobiographical Writings*, translated by Peter Demetz. New York: Schocken, 1986.

Benjamin, Walter. *Reflections: Essays, Aphorisms, Autobiographical Writings*. Translated by Peter Demetz. New York: Schocken, 1986.

Bernes, Jasper. "Documents." In *Starsdown*, 82. Berkeley: Small Press Distribution, 2007.

Blum, Hester, Mary Eyring, Iping Liang, and Brian Roberts Russell. "Introduc- tion to the Special Forum on Archipelagoes/Oceans/American Visuality." *Journal of Transnational American Studies* 10 (2019): 5–21.

Boehmer, Eileke. *Colonial and Postcolonial Literature: Migrant Metaphors*. Oxford: Oxford University Press, 2005.

Bourne, Randolph S. "Trans-national America." *Atlantic Monthly* 118, no. 1 (July 1916): 86–97. http://www.theatlantic.com/past/issues/16jul/bourne .htm.

Brathwaite, Kamau. *Middle Passages*. New York: New Directions, 1994.

Brautigan, Richard. *Trout Fishing in America*. New York: Dell, 1967.

Brechin, Gray. *Imperial San Francisco: Urban Power, Earthly Ruin*. Berkeley: University of California Press, 1999.

Brislin, Tom. "On Berlin." *Honolulu Advertiser*, November 17, 1999.

Brown, Norman O. "From Politics to Metapolitics." In Dale Pendell, *Walking with Nobby: Conversations with Norman O. Brown*, 201–14. San Francisco: Mercury House, 2008.

Brown, Norman O. *Love's Body*. Berkeley: University of California Press, 1966.

Buell, Lawrence. *Writing for an Endangered World: Literature, Culture, and Environment in the US and Beyond*. Cambridge, MA: Harvard University Press, 2001.

Burke, Kenneth. *A Grammar of Motives*. Berkeley: University of California Press, 1969.

Burlingame, Burl. "Wanderers Find Poetry in the Environment." *Honolulu Star-Bulletin*, March 1, 2000. http://archives.starbulletin.com/2000/03/01/features/story2.html.

Carson, Rachel L. *The Sea around Us*. London: Oxford University Press, 1953.

Castan Broto, Vanessa, and Enora Robin. "Climate Urbanism as Critical Urban Theory." *Urban Geography* 42, no. 6 (2020). https://doi.org/10.1080/02723638.2020.1850617.

Chan, Tara Francis. "China Is Not Even Pretending Anymore in the South China Sea: It Put 400 Buildings on One of the Disputed Islands." *Business Insider*, May 24, 2018. http://www.businessinsider.com/china-400-buildings-subi-reef-south-china-sea-2018-5.

Channer, Colin. "Spumante." *New Yorker*, November 9, 2020. https://www.newyorker.com/magazine/2020/11/16/spumante.

Cheah, Pheng. "What Is a World? On Literature as a World-Making Activity." *Daedalus* 137, no. 3 (Summer 2008): 26–38.

Cheah, Pheng. *What Is a World? On Postcolonial Literature as World Literature*. Durham, NC: Duke University Press, 2016.

Cheah, Pheng. "World against Globe: Toward a Normative Conception of World Literature." *New Literary History* 45 (2014): 303–29.

Chen, Kuan-hsing. *Asia as Method: Toward Deimperialization*. Durham, NC: Duke University Press, 2010.

Chen, Kuan-hsing. "Taiwan as Club 51: On the Culture of US Imperialism." In *The Worlding Project: Doing Cultural Studies in the Era of Globalization*, edited by Rob Wilson and Christopher Leigh Connery, 109–31. Berkeley, CA: North Atlantic and New Pacific, 2007.

Ch'ien, Evelyn Nien-Ming. *Weird English*. Cambridge, MA: Harvard University Press, 2004.

Child, Ben. "Pacific Rim Designed to Advance US Cultural Domination of China." *Guardian*, August 28, 2013. http://www.theguardian.com/film/2013/aug/28/pacific-rim-us-cultural-domination-china.

Chin, Tamara. *Savage Exchange: Han Imperialism, Chinese Literary Style, and the Economic Imagination*. Cambridge. MA: Harvard University Press, 2014.

Cho, Yu-Fang. "Remembering Lucky Dragon, Re-membering Bikini: Worlding the Anthropocene through Transpacific Nuclear Modernity." *Cultural Studies* 33 (2018): 122–46.

Chong, Terence. "Bukit Brown Municipal Cemetery: Contesting Imaginations of the Good Life in Singapore." In *Worlding Multiculturalisms: The Politics of Inter-Asian Dwelling*, edited by Daniel P. S. Goh, 161–82. London: Routledge, 2015.

Chou, Shiuhhuah Serena. "Agrarianism in the City: Urban Agriculture and the Anthropocene Futurity." *Concentric: Literary and Cultural Studies* 43 (March 2017): 51–69.

Chou, Shiuhhuah Serena. "Claiming the Sacred: Indigenous Knowledge, Spiritual Ecology, and the Emergence of Eco-cosmopolitanism." *Cultura* 12, no. 1 (2015): 71–84.

Chou, Shiuhhuah Serena, Soyoung Kim, and Rob Wilson, eds. *Geo-spatiality in Asian and Oceanic Literature and Culture: Worlding Asia in the Anthropocene*. London: Palgrave Macmillan, 2022.

Choy, Catherine Ceniza. *Empire of Care: Nursing and Migration in Filipino American History*. Durham, NC: Duke University Press, 2013.

Christensen, Jon, Jan Goggans, and Ursula K. Heise. "The True Literature of California Is Science Fiction" (interview with Kim Stanley Robinson). *Boom*, January 30, 2014. http://www.gizmodo.com.au/2014/01/the-true-literature -of-california-is-science-fiction.

Chua Beng Huat. "Singapore as Model: Planning Innovations, Knowledge Experts." In *Worlding Cities: Asian Experiments and the Art of Being Global*, edited by Ananya Roy and Aihwa Ong, 27–54. Chichester, UK: Wiley-Blackwell, 2011.

Chuihua, Judy Chung, Jeffrey Inaba, Rem Koolhaas, and Sze Tsung Leong, eds. *Project on the City I: Great Leap Forward*. Cologne, Germany: Taschen, 2001.

Chyung, Eun-ju, and Si-soo Park. "Rich Marine Species Vanish with Climate Change." *Korea Times*, July 24, 2017. http://www.koreatimes.co.kr/www /nation/2017/07/371_233516.html.

Clifford, James. "Indigenous Artiulations." In *The Worlding Project: Doing Cultural Studies in the Era of Globalization*, edited by Rob Wilson and Christopher Leigh Connery, 13–38. Berkeley, CA: North Atlantic and New Pacific, 2007.

Clifford, James. *Returns: Becoming Indigenous in the Twenty-first Century*. Cambridge, MA: Harvard University Press, 2013.

Clifford, James. *Routes: Travel and Translation in the Late Twentieth Century*. Cambridge, MA: Harvard University Press, 1997.

Clover, Joshua. *Red Epic*. Oakland, CA: Commune, 2015.

Clover, Joshua. *Riot. Strike. Riot: The New Era of Uprisings*. London: Verso, 2016.

Cocteau, Jean. *Round the World Again in Eighty Days: Tour du Monde en 80 Jours*. London: Tauris Parke, 2000.

Coffman, Tom. *Nation Within: The History of the American Occupation of Hawai'i*. Durham, NC: Duke University Press, 2016.

Cohen, Jeffrey Jerome, ed. *Prismatic Ecology: Ecotheory beyond Green*. Minneapolis: University of Minnesota Press, 2013.

Comer, Krista. *Surfer Girls in the New World Order*. Durham, NC: Duke University Press, 2010.

Connery, Christopher L. "Better City, Better Life." *boundary 2*, vol. 38, no. 2 (2011): 207–27.

Connery, Christopher, L. "Ideologies of Land and Sea: Alfred Thayer Mahan, Carl Schmitt, and the Shaping of Global Myth Elements." *boundary 2*, vol. 28, no. 2 (2001): 173–201.

Connery, Christopher L. "The Oceanic Feeling and the Regional Imaginary." In *Global/Local: Cultural Production and the Transnational Imaginary*, edited by Rob Wilson and Wimal Dissanayake, 284–311. Durham, NC: Duke University Press, 1996.

Connery, Christopher L. "Pacific Rim Discourse." In *Asia/Pacific as Space of Cultural Production*, edited by Rob Wilson and Arif Dirlik, 30–56. Durham, NC: Duke University Press, 1995.

Connery, Christopher L. "Sea Power." PMLA 125 (2010): 685–92.

Connery, Christopher L. "There Was No More Sea: The Supercession of the Ocean from the Bible to Cyberspace." *Journal of World Geography* 32 (2006): 495–511.

Copp, Tara. "Carter to Visit Beijing, Increase US-China Security Exercises." *Stars and Stripes*, June 4, 2016. http://www.stripes.com/news/pacific/carter-to-visit-beijing-increase-us-china-security-exercises-1.413106.

Cumings, Bruce. *Dominion from Sea to Sea: Pacific Ascendancy and American Power*. New Haven, CT: Yale University Press, 2009.

Dash, J. Michael. "The Stranger by the Shore: The Archipelization of Caliban in Antillean Theatre." In *Archipelagic American Studies*, edited by Brian Russell Roberts and Michelle Ann Stephens, 356–72. Durham, NC: Duke University Press, 2017.

Davis, Alan E. "A Preliminary List of Animal Names in the Chuuk District, with Notes on Plant Names." *Micronesica* 31, no. 1 (1999): 1–245. https://micronesica.org/sites/default/files/1_davis_chuuk_lexicon.pdf.

Deleuze, Gilles. *Essays Critical and Clinical*. Translated by Daniel W. Smith and Michael A. Greco. Minneapolis: University of Minnesota Press, 1997.

Deleuze, Gilles, and Félix Guattari. *Anti-Oedipus: Capitalism and Schizophrenia*. Translated by Robert Hurley, Mark Seem, and Helen R. Lane. Minneapolis: University of Minnesota Press, 1983.

Deleuze, Gilles, and Félix Guattari. *Kafka: Toward a Minor Literature*. Translated by Dana Polan. Minneapolis: University of Minnesota Press, 1986.

Deleuze, Gilles, with Clare Parnet. *Dialogues*. Translated by Hugh Tomlinson and Barbara Habberjam. New York: Columbia University Press, 1987.

DeLoughrey, Elizabeth M. *Allegories of the Anthropocene*. Durham, NC: Duke University Press, 2019.

DeLoughrey, Elizabeth M. "Heavy Waters: Waste and Atlantic Modernity." PMLA 125 (2010): 703–12.

DeLoughrey, Elizabeth M. "Ordinary Futures: Interspecies Worldings in the Anthropocene." In *Global Ecologies and the Environmental Humanities: Postcolonial Approaches*, edited by Elizabeth DeLoughrey, Jil Didur, and Anthony Carrigan, 352–72. London: Routledge, 2015.

DeLoughrey, Elizabeth M., and George B. Handley, eds. *Postcolonial Ecologies: Literature of the Environment*. New York: Oxford University Press, 2011.

Demick, Barbara. "Malaysia Plane: Confronting Searchers Is an Ocean Full of Garbage." *Los Angeles Times*, March 30, 2014. https://www.latimes.com/world /worldnow/la-fg-wn-malaysia-plane-search-garbage-20140330-story.html.

Demos, T. J. "Welcome to the Anthropocene!" *Foto-museum* (blog), May 5, 2015. http://blog.fotomuseum.ch/2015/05/i-welcome-to-the-anthropocene.

D'haen, Theo. "Worlding World Literature." *Recherches Littéraires / Literary Research* 32 (2016): 7–23.

"Diane Di Prima's 'Rant.'" *Modern Americans* (blog), April 12, 2006. http:// modampo.blogspot.com/2006/04/diane-di-primas-rant.html.

Dimock, Wai Chee. *Through Other Continents: American Literature across Deep Time*. Princeton, NJ: Princeton University Press, 2006.

Dimock, Wai Chee. *Weak Planet: Literature and Assisted Survival*. Chicago: University of Chicago Press, 2020.

Dirlik, Arif. "Asia Pacific Studies in an Age of Global Modernity." *Inter-Asia Cultural Studies* 6 (2005): 158–70.

Dirlik, Arif, ed. *What Is in a Rim? Critical Perspectives on the Pacific Region Idea*. Lanham, MD: Rowman and Littlefield, 1998.

Dissanayake, Wimal, and Rob Wilson, eds. *Global/Local: Cultural Production and the Transnational Imaginary*. Durham, NC: Duke University Press, 1996.

Duras, Marguerite. *Hiroshima Mon Amour*. Translated by Richard Seaver. New York: Grove, 1961.

Dynes, Robert. "California's Competitiveness Starts with Research Universities." *San Francisco Chronicle*, April 16, 2004, B9.

Economic Times. "US Pacific Command Renamed as US Indo-Pacific Command." May 31, 2018. https://economictimes.indiatimes.com/news/defence /us-pacific-command-renamed-as-us-indo-pacific-command/article show/64398189.cms.

Economist. "Make Law, Not War: How to Solve Spats over Sea Borders." August 25, 2012. http://www.economist.com/node/21560849.

Economist. "Too Small an Ocean." September 8, 2012. https://www.economist.com /china/2012/09/08/too-small-an-ocean.

Eisen-Martin, Tongo. *Heaven Is All Goodbyes*. San Francisco: City Lights, 2017.

Eisen-Martin, Tongo. *someone's already dead*. Oakland, CA: Bootstrap, 2015.

Emerson, Nathaniel Bright. *Unwritten Literature of Hawai'i: The Sacred Songs of the Hula*. Washington, DC: Smithsonian Institute, 1909.

Eperjesi, John. "Basing the Pacific: Exceptional Spaces of the Wilkes Exploring Expedition, 1838–1842." *Amerasia Journal* 37 (2011): 1–17.

Estok, Simon C., and Shiuhhuah Serena Chou, eds. "The City and the Anthropocene." Foreword to the special issue of *Concentric: Literary and Cultural
Studies* 43, no. 1 (March 2017): 3–11. https://doi.org/10.6240/concentric.lit
.2017.43.1.01.

Ferguson, Frances. "The Nuclear Sublime." *Diacritics* 14 (1984): 4–10.

Fisher-Wirth, Ann, and Laura-Gray Street, eds. *The Ecopoetry Anthology*. San
Antonio: Trinity University Press, 2013.

Foucault, Michel. "The Refugee Problem Is a Presage of the Great Migrations
of the Twenty-First Century." *Shûkan Posoto*, August 1979. Translated
by Colin Gordon for *Open Democracy*, November 13, 2015. https://www
.opendemocracy.net/michel-foucault/refugee-problem-is-presage-of-great
-migrations-of-twenty-first-century.

Fussell, Paul. *Thank God for the Atom Bomb and Other Essays*. New York: Summit,
1988.

Garikipati, Rambabu. "APEC Provides Platform for Free Trade, Investment." *Korea
Herald*, November 11, 2004, 3.

Ghosh, Amitav. *The Great Derangement: Climate Change and the Unthinkable*.
Chicago: University of Chicago Press, 2016.

Ghosh, Rajan. "Albatross Unbound: Worlding the Plastic Sea." In *Worlding Asia:
Asian/Pacific/American/Planetary Convergences*, edited by Shiuhhuah
Serena Chou, Soyoung Kim, and Rob Wilson, 247–65. London: Palgrave
Macmillan, 2021.

Gillis, John R. "The Blue Humanities." *Humanities* 34, no. 3, May–June 2013.
https://www.neh.gov/humanities/2013/mayjune/feature/the-blue
-humanities.

Gillis, John R. *The Human Shore: Seacoasts in History*. Chicago: University of
Chicago Press, 2012.

Gillman, Susan. *American Mediterraneans: A Study in Geography, History, and
Race*. Chicago: University of Chicago Press, 2022.

Gilroy, Paul. *After Empire: Melancholia or Convivial Culture?* Abingdon, UK:
Routledge, 2006.

Gilroy, Paul. *Postcolonial Melancholia*. New York: Columbia University Press, 2006.

Ginsberg, Allen. *Collected Poems, 1947–1980*. New York: Harper and Row, 1984.

Ginsberg, Allen. *Howl and Other Poems*. San Francisco: City Lights, 1956.

Ginsberg, Allen. *White Shroud: Poems 1980–1985*. New York: Harper, 1986.

Glissant, Édouard. *Poetics of Relation*. Translated by Betsy Wing. Ann Arbor:
University of Michigan Press, 1997.

Gluck, Carol, and Anna Lowenhaupt Tsing, eds. *Words in Motion: Toward a
Global Lexicon*. Durham, NC: Duke University Press, 2009.

Goh, Daniel P. S., ed. *Worlding Multiculturalisms: The Politics of Inter-Asian
Dwelling*. London: Routledge, 2015.

Goldman, Michael. "Speculating on the Next World City." In *Worlding Cities:
Asian Experiments and the Art of Being Global*, edited by Ananya Roy and
Aihwa Ong, 229–59. Chichester, UK: Wiley-Blackwell, 2011.

Gonick, Sophie. "Global Cities and Their Discontents: Saskia Sassen and Teresia Caldeira in Conversation." *Public Books*, June 13, 2019. https://www.publicbooks.org/global-cities-and-their-discontents-saskia-sassen-and-teresa-caldeira.

Greene, Andy. "New Steve Jobs Bio [by Walter Isaacson] Reveals Details of His Relationship with Bob Dylan and Bono." *Rolling Stone*, October 24, 2011. https://www.rollingstone.com/music/news/new-steve-jobs-bio-reveals-details-of-his-relationships-with-bob-dylan-bono-20111024.

Guattari, Félix. *The Three Ecologies*. Translated by Gary Genosko. London: Athlone, 2000.

Hadot, Pierre. "Conversion." In *Encyclopaedia Universalis* 4:979–81. Paris: Encyclopaedia Universalis France, 1968.

Hagedorn, Jessica. "The Exile Within / The Question of Identity." In *The State of Asian America: Activism and Resistance in the 1990s*, edited by Karin Aguilar–San Juan, 173–82. Boston: South Point, 1994.

Haines, Chad. "Cracks in the Façade: Landscapes of Hope and Desire in Dubai." In *Worlding Cities: Asian Experiments and the Art of Being Global*, edited by Ananya Roy and Aihwa Ong, 160–81. Chichester, UK: Wiley-Blackwell, 2011.

Haraway, Donna. "Anthropocene, Capitalocene, Plantationocene, Chthulucene / Making Kin." *Environmental Humanities* 6 (2015): 159–65.

Haraway, Donna. "A Giant Bumptious Litter: Donna Haraway on Truth, Technology, and Resisting Extinction." *Logic(s)*, December 7, 2019. https://logicmag.io/nature/a-giant-bumptious-litter.

Haraway, Donna. *Staying with the Trouble: Making Kin in the Chthulucene*. Durham, NC: Duke University Press, 2016.

Hardt, Michael, and Antonio Negri. "Marx's Mole Is Dead!" In *The Communist Manifesto: Norton Critical Edition*, edited by Frederic L. Bender, 209–25. New York: W. W. Norton, 2012.

Harvey, David. *Seventeen Contradictions and the End of Capitalism*. New York: Oxford University Press, 2014.

Hau'ofa, Epeli. *Kisses in the Nederends*. Honolulu: University of Hawai'i Press, 1995.

Hau'ofa, Epeli. *Tales of the Tikongs*. Honolulu: University of Hawai'i Press, 1994.

Hau'ofa, Epeli. *We Are the Ocean: Selected Works*. Honolulu: University of Hawai'i Press, 2008.

Hayot, Eric. *On Literary Worlds*. New York: Oxford University Press, 2012.

Hayton, Bill. *The South China Sea: The Struggle for Power in Asia*. New Haven, CT: Yale University Press, 2014.

Helmreich, Stefan. *Alien Ocean: Anthropological Voyages in Microbial Seas*. Berkeley: University of California Press, 2009.

Hendrix, Jerry. "America's Future Is at Sea." *Atlantic Monthly*, vol. 331, April 2023, 51–57.

Hereniko, Vilsoni, and Rob Wilson, eds. *Inside Out: Literature, Cultural Politics, and Identity in the New Pacific*. Lanham, MD: Rowman and Littlefield, 1999.

Hillman, James. "Wars, Arms, Rams, Mars." In *Mythic Figures*. Uniform Edition of the Writings of James Hillman, vol. 6. Thompson, CT: Spring, 2007.

Hjorth, Larissa, and Dean Chan, eds. *Gaming Culture and Place in Asia-Pacific.* London: Routledge, 2009.

Ho, Vincent. "Casino Multiculturalism and the Reinvention of Heritage in Macao." In *Worlding Multiculturalisms: The Politics of Inter-Asian Dwelling*, edited by Daniel P. S. Goh, 129–42. London: Routledge, 2015.

Hockenos, Paul. "Zero Hour: The First Days of New Berlin." *Boston Review*, November 4, 2014.

"Hong Kong's Taste for Seafood Putting Oceans in Danger." *Japan Times*, August 12, 2015. http://www.japantimes.co.jp/news/2015/08/12/asia-pacific /science-health-asia-pacific/hong-kongs-taste-for-seafood-putting-oceans -in-danger/#.VcycIhd3ib5.

Hopkins, Gerard Manley. "The Blessed Virgin Compared to the Air We Breathe." *Official Gerard Manley Hopkins Website*, December 10, 2018. https:// hopkinspoetry.com/poem/the-blessed-virgin.

Hsiung, James C. "Sea Power, Law of the Sea, and Sino-Japanese East China Sea 'Resource War.'" In *China and Japan at Odds*, 133–53. New York: Palgrave Macmillan, 2007. https://doi.org/10.1057/9780230607118_8.

Hsu, Hua. *A Floating Chinaman: Fantasy and Failure across the Pacific.* Cambridge, MA: Harvard University Press, 2016.

Huang, Hsinya, ed. "Islands: Narratives, Aesthetics, Ecologies." Special issue of *Comparative Literature Association of Taiwan Newsletter*, no. 17 (June 2016).

Huang, Hsinya. "Performing Archipelagic Identities in Bill Reid, Robert Sullivan and Shyman Raporgan." In *Archipelagic American Studies*, edited by Brian Russell Roberts and Michelle Ann Stephens, 281–301. Durham, NC: Duke University Press, 2017.

Huang, Hsinya. "Towards Transnational Native American Literary Studies." *Comparative Literature and Culture* 13 (2011). http://docs.lib.purdue.edu /clcweb/vol13/iss2/6.

Huang, Hsinya, and Chia-hua Lin, eds. *Pacific Literatures as World Literature.* New York: Bloomsbury, 2023.

Huang, Peter I-Min. *Linda Hogan and Contemporary Taiwanese Writers: An Ecocritical Study of Indigeneities and Environment.* Lanham, MD: Lexington, 2016.

Hu-DeHart, Evelyn, ed. *Across the Pacific: Asian Americans and Globalization.* Philadelphia: Temple University Press, 1999.

Hull, Christopher. "'Hiroshima: Out of the Ashes' Takes Point of View of Victims." *Honolulu Star-Bulletin and Advertiser*, August 5, 1990.

Huyssen, Andreas. *After the Great Divide: Modernism, Mass Culture, Postmodernism.* Bloomington: Indiana University Press, 1986.

Ibuse, Masuji. *Black Rain.* Translated by John Bester. New York: Bantam, 1985.

Ibuse, Masuji. "The Crazy Iris." In *The Crazy Iris and Other Stories of the Atomic Age*, edited by Kenzaburō Ōe, 17–35. New York: Grove, 1985.

Igler, David. *The Great Ocean: Pacific Worlds from Captain Cook to the Gold Rush.* New York: Oxford University Press, 2013.

Imada, Adria L. *Aloha America: Hula Circuits through the US Empire*. Durham, NC: Duke University Press, 2012.

Inada, Lawson. "Shrinking the Pacific." In *Asia/Pacific as Space of Cultural Production*, edited by Rob Wilson and Arif Dirlik, 80–81. Durham, NC: Duke University Press, 1995.

Iwabuchi, Koichi. *Recentering Globalization: Popular Culture and Japanese Transnationalism*. Durham, NC: Duke University Press, 2002.

Jameson, Fredric. *The Cultural Turn: Selected Writings on the Postmodern, 1983–1998*. London: Verso, 1998.

Jameson, Fredric. "New Literary History after the End of the New." *New Literary History* 39 (2008): 375–87.

Jameson, Fredric. *A Singular Modernity: Essay on the Ontology of the Present*. London: Verso, 2002.

Jetnil-Kijiner, Kathy. *Iep Jaltok: Poems from a Marshallese Daughter*. Tucson: University of Arizona Press, 2017.

Johnson, Chalmers. *Blowback: The Costs and Consequences of American Empire*. New York: Henry Holt, 2000.

Johnson, Chalmers. *The Sorrows of Empire: Militarism Secrecy, and the End of the Republic*. New York: Henry Holt, 2004.

Juvan, Marko. "Worlding Literatures between Dialogue and Hegemony." *Comparative Literature and Culture* 15, no. 5 (2013). https://doi.org/10.7771/1481-374.2343.

Kaiser, Birgit Mara. "Worlding Comp Lit: Diffractive Reading with Barad, Glissant, and Nancy." *Parallax* 20 (2014): 247–87.

Kalākaua, King David. *The Legends and Myths of Hawaii*. Honolulu: Mutual, 1990.

Kamakau, Samuel M. *Ruling Chiefs of Hawaii*. Rev. ed. Honolulu: Kamehameha Schools Press, 1992.

Kaplan, Robert D. *Asia's Cauldron: The South China Sea and the End of a Stable Pacific*. New York: Random House, 2014.

Kaplan, Robert D. "Geography Strikes Back." *Wall Street Journal*, September 7, 2012. http://online.wsj.com/article/SB10000872396390443819404577635332556005436.html.

Karatani, Kojin. *Origins of Modern Japanese Literature*. Translated by Brett De Bary. Durham, NC: Duke University Press, 1993.

Kaufman, Bob. *Cranial Guitar: Selected Poems by Bob Kaufman*. Minneapolis: Coffee House, 1995.

Kaufman, Bob. *Solitudes Crowded with Loneliness*. New York: New Directions, 1965.

Keown, Michelle, Andrew Taylor, and Mandy Treagus, eds. *Anglo-American Imperialism and the Pacific: Discourses of Encounter*. London: Routledge, 2019.

Kerouac, Jack, *Big Sur*. New York: Bantam, 1962.

Kerouac, Jack. *Desolation Angels*. New York: Riverhead, 1995.

Kerouac, Jack. *The Dharma Bums*. New York: Penguin, 1976.

Kerouac, Jack. *On the Road: The Original Scroll*. Edited by Howard Cunnell. New York: Viking, 2007.

Kerouac, Jack. *The Portable Jack Kerouac*. Edited by Ann Charters. New York: Penguin, 2007.

Kerouac, Jack. *The Sea Is My Brother: The Lost Novel*. Edited by Dawn Ward. London: Penguin, 2011.

Kilvert, Nick. "Great Pacific Garbage Patch Plastic Removal System Could Become 'World's Biggest Piece of Marine Debris,' Critics Say." ABC *Science News*, May 7, 2018. http://www.abc.net.au/news/science/2018-05-08/great -pacific-garbage-patch-plastic-ocean-cleanup-boyan-slat/9714246.

Kim, Annette M. "Satellite Images Can Harm the Poorest Citizens." *Atlantic*, June 5, 2018. https://www.theatlantic.com/technology/archive/2018/06 /satellite-images-can-harm-the-poorest-citizens/561920.

Kim, Jinah. *Postcolonial Grief: The Afterlives of the Pacific Wars in the Americas*. Durham, NC: Duke University Press, 2019.

Kim, Sung Kyung. "Shopping Mall as Dwelling-Place: Multiculturalism and the Spatial Struggle over Time Square, Seoul." In *Worlding Multiculturalisms: The Politics of Inter-Asian Dwelling*, edited by Daniel P. S. Goh, 144–60. London: Routledge, 2015.

King, Tiffany Lethabo. *The Black Shoals: Offshore Formations of Black and Native Studies*. Durham, NC: Duke University Press, 2019.

Kingston, Maxine Hong. *The Fifth Book of Peace*. New York: Vintage, 2003.

Kingston, Maxine Hong. *Tripmaster Monkey*. New York: Vintage, 1990.

Kirksey, Eben. "Queer Love, Gender Bending Bacteria, and Life after the Anthropocene." *Theory, Culture, and Society* 36, no. 6 (June 3, 2018): 1–23. http:// journals.sagepub.com/eprint/dVTVmExbCUWqXYdYGzw2/full.

Kissinger, Henry. *On China*. New York: Penguin, 2011.

Klein, Naomi. *This Changes Everything: Capitalism versus the Climate*. New York: Simon and Schuster, 2014.

Knott, Bill. *The Unsubscriber: Poems*. New York: Farrar, Straus and Giroux, 2004.

Kuo, Mercy A. "One Belt, One Road: A Convergence of Civilizations? Insights from Tamara Chin." *Diplomat*, May 24, 2017. https://thediplomat.com /2017/05/one-belt-one-road-a-convergence-of-civilizations.

Lapham, Louis, ed. "The Sea." Special issue of *Lapham's Quarterly* 6 (Summer 2013).

Lapin, Andrew. "'Chasing Coral': Documentary Vividly Chronicles a Growing Threat to Oceans." National Public Radio, July 13, 2017. http://www.npr.org /2017/07/13/536644965/chasing-coral-documentary-vividly-chronicles-a -growing-threat-to-oceans.

Lee, Erika. *The Making of Asian America: A History*. New York: Simon and Schuster, 2015.

Lee, Zack. "Eco-cities as an Assemblage of *Worlding* Practices." *International Journal of Built Environment and Sustainability* 2 (2015): 183–91.

Leonard, Annie. "Our Plastic Pollution Crisis Is Too Big for Recycling to Fix." *Guardian*, June 9, 2018. https://www.theguardian.com/commentisfree /2018/jun/09/recycling-plastic-crisis-oceans-pollution-corporate -responsibility.

Lepawsky, Josh. *Reassembling Rubbish: Worlding Electronic Waste.* Cambridge, MA: MIT Press, 2018.

Leslie, Esther, "Walter Benjamin: The Refugee and Migrant." *Verso* (blog), October 14, 2015. http://www.versobooks.com/blogs/2283-walter-benjamin-the -refugee-and-migrant-by-esther-leslie.

Liboiron, Max. *Pollution Is Colonialism.* Durham, NC: Duke University Press, 2021.

Liliʻuokalani, Queen. *Hawaii's Story by Hawaii's Queen.* Honolulu: Mutual, 1990.

Litzinger, Ralph. "The Labor Question in China: Apple and Beyond." *South Atlantic Quarterly* 112, no. 1 (Winter 2013): 172–78.

Lowe, Lisa. *The Intimacies of Four Continents.* Durham, NC: Duke University Press, 2015.

Lowrie, Annie. "Why the Phrase 'Late Capitalism' Is Suddenly Everywhere." *Atlantic,* May 1, 2017. https://www.theatlantic.com/business/archive/2017/05 /late-capitalism/524943.

Lye, Colleen. *America's Asia: Racial Form and American Literature, 1893–1945.* Princeton, NJ: Princeton University Press, 2004.

Lynas, Mark. *The God Species: How the Planet Can Survive the Age of Humans.* New York: National Geographic, 2011.

Lyons, Laura. "American Pacific Culture and Theory." *Years' Work in Critical and Cultural Theory* 6 (1996): 315–24.

MacFarlane, Key. "Marshland: Hamburg G20 and the Return of the Hanseatic League." *Society and Space,* July 18, 2018. https://societyandspace.org /author/key-macfarlane.

MacFarlane, Key, and Katharyne Mitchell. "Hamburg's Spaces of Danger: Race, Violence, and Memory in a Contemporary Global City." *International Journal of Urban and Regional Research* 43 (2019): 816–32.

Mackey, Nathaniel. *Discrepant Engagement: Dissonance, Cross-culturality, and Experimental Writing.* Tuscaloosa: University of Alabama Press, 2000.

Mackey, Nathaniel. *Double Trio.* New York: New Directions, 2021.

Mackey, Nathaniel. "Song of the Andoumboulou: 300." *boundary* 2, vol. 49, no. 3 (2022): 55–64. https://doi.org/10.1215/01903659-9789682.

Mackey, Nathaniel. *Splay Anthem.* New York: New Directions, 2006.

Mardorossian, Carine M. "From Literature of Exile to Migrant Literature." *Modern Language Studies* 32 (2002): 15–33.

Marran, Christine L. "The Planetary." *boundary* 2, vol. 46, no. 3 (2019): 181–98.

Marx, Karl. *The Communist Manifesto: Norton Critical Edition.* Edited by Frederic L. Bender. New York: W. W. Norton, 2013.

McDougall, Brandy Nalani. *Finding Meaning: Kaona and Contemporary Hawaiian Literature.* Tucson: University of Arizona Press, 2016.

McDougall, Brandy Nalani. *The Salt Wind, Ka Makani Paʻakai.* Honolulu: Kuleana Oʻiwi, 2008.

McHugh, Paul. "Poet Ferlinghetti Chased Subs in World War II." *San Francisco Chronicle,* January 12, 2012. https://www.sfgate.com/news/article/Poet -Ferlinghetti-chased-subs-in-WWII-2484205.php.

Mearsheimer, John J. "The Gathering Storm: China's Challenge to US Power in Asia." *Chinese Journal of International Politics* 3 (2010): 381–96.

Meltzer, David. "Interview with Kenneth Rexroth, Summer 1969." In *The San Francisco Poets*, edited by David Meltzer. New York: Ballantine, 1971. https://www.bopsecrets.org/rexroth/meltzer.htm.

Meltzer, David, ed. *The San Francisco Poets*. New York: Ballantine, 1971.

Melville, Herman. *Moby-Dick*. New York: Dover, 2015.

Mentz, Steve. *At the Bottom of Shakespeare's Ocean*. London: Continuum, 2009.

Mentz, Steve. *An Introduction to the Blue Humanities*. London: Routledge, 2023.

Mentz, Steve. *Ocean*. New York: Bloomsbury, 2020.

Mentz, Steve. *Shipwreck Modernity: Ecologies of Globalization, 1550–1719*. Minneapolis: University of Minnesota Press, 2015.

Mirzoeff, Nick. "The Politics of Seeing with the Global City." *Hyperallergic*, May 29, 2018. https://hyperallergic.com/444073/the-politics-of-seeing-within-the-global-city.

Mitchell, Katharyne. "Freedom, Faith, and Humanitarian Governance: The Spatial Politics of Church Asylum in Europe." *Space and Polity* 21, no. 3 (2017): 269–88. https://doi.org/10.1080/13562576.2017.1380883.

Miyoshi, Masao. *Trespasses: Selected Writings*. Edited by Eric Cazdyn. Durham, NC: Duke University Press, 2010.

Miyoshi, Masao, "Turn to the Planet: Literature, Diversity, and Totality," *Comparative Literature* 53, no. 4 (2001): 283–97.

Moore, Charles, with Cassandra Phillips. *Plastic Ocean: How a Sea Captain's Chance Discovery Launched a Determined Quest to Save the Oceans*. New York: Avery, 2011.

Morris, Meaghan. "Politics Now (Anxieties of a Petit-bourgeois Intellectual)." *Framework: The Journal of Cinema and Media*, no. 32–33 (1986): 4–19. https://www.jstor.org/stable/44111110.

Morton, Timothy. *Hyperobjects: Philosophy and Ecology at the End of the World*. Minneapolis: University of Minnesota Press, 2013.

Nail, Thomas. "Alain Badiou and the *Sans-Papiers*." *Angelaki* 20 (2015): 109–30.

Nail, Thomas. *The Figure of the Migrant*. Stanford, CA: Stanford University Press, 2015.

Nancy, Jean-Luc. *The Creation of the World or Globalization*. Translated by Francois Raffoul and David Pettigrew. Albany: State University of New York Press, 2007.

New York Times. "U.S., China Clash on Key Issues." September 5, 2012.

Nguyen, Viet Than. *The Displaced: Refugee Writers on Refugee Life*. New York: Harry N. Abrams, 2018.

Nguyen, Viet Than. *Nothing Ever Dies: Vietnam and the Memory of War*. Cambridge: Harvard University Press, 2017.

Nguyen, Viet Than. *The Refugees*. New York: Grove, 2018.

Nguyen, Viet Than. *The Sympathizer*. New York: Grove, 2016.

Nguyen, Viet Than, and Janet Alison Hoskins, eds. *Transpacific Studies: Framing an Emergent Field*. Honolulu: University of Hawai'i Press, 2014.

Nichols, Wallace J. *Blue Mind: The Surprising Science That Shows How Being near, in, on, or under Water Can Make You Happier, Healthier, More Connected, and Better at What You Do*. New York: Little, Brown, 2014.

Nixon, Rob. "The Great Acceleration and the Great Divergence: Vulnerability during the Anthropocene." *Profession*, March 2014. https://profession.mla .org/the-great-acceleration-and-the-great-divergence-vulnerability-in-the -anthropocene/.

Nixon, Rob. *Slow Violence and the Environmentalism of the Poor*. Cambridge, MA: Harvard University Press, 2011.

"Nuclear Doomsday Clock, The." *Bulletin of the Atomic Scientists*, January 23, 2024. https://thebulletin.org/doomsday-clock/current-time/nuclear-risk.

Nye, David E. *American Technological Sublime*. Cambridge, MA: MIT Press, 1994.

Ōe, Kenzaburō, ed. *The Crazy Iris and Other Stories of the Atomic Aftermath*. New York: Grove, 1985.

Ōe, Kenzaburō. *Hiroshima Notes*. Translated by Toshi Yonezawa. Tokyo: YMCA Press, 1981.

Ōe, Kenzaburō. "Japan's Dual Identity: A Writer's Dilemma." In *Postmodernism and Japan*, edited by Masao Miyoshi and Harry D. Harootunian, 191–201. Durham, NC: Duke University Press, 1989.

Ōe, Kenzaburō. "The Myth of My Own Village" ("An Interview with Kenzaburō Ōe: The Myth of My Own Village," with Steve Bradbury, Joel Cohn, and Rob Wilson). *Mānoa* 6 (1994): 135–44.

Ōe, Kenzaburō. *A Personal Matter*. Translated by John Nathan. New York: Grove, 1969.

Ōe, Kenzaburō. *The Silent Cry*. Translated by John Bester. New York: Kodansha International, 1974.

Okeowo, Alexis. "The Writing Life of a Young, Prolific Poet [Warshan Shire]." *New Yorker*, October 21, 2015. http://www.newyorker.com/culture/cultural -comment/the-writing-life-of-a-young-prolific-poet-warsan-shire.

Olson, Charles. *Call Me Ishmael*. Baltimore: Johns Hopkins University Press, 1997.

Ong, Aihwa. "Introduction: Worlding Cities, or the Art of Being Global." In *Worlding Cities: Asian Experiments and the Art of Being Global*, edited by Ananya Roy and Aihwa Ong, 1–26. Chichester, UK: Wiley-Blackwell, 2011.

Pai, Hsiao-Hung. *Scattered Sands: The Story of China's Rural Migrants*. London: Verso, 2012.

Palmer, Jane. "Junk Accumulating on Monterey Bay Ocean Floors: Scientists Find Increasing Levels of Debris in the Deep Sea." *Santa Cruz Sentinel*, February 2, 2010.

Park, Josephine. *Apparitions of Asia: Modernist Form and Asian American Poetics*. London: Oxford University Press, 2008.

Pease, Donald E. "Hiroshima, the Vietnam Veterans War Memorial and the Gulf War: Post-national Spectacles." In *Cultures of United States Imperialism*,

edited by Donald E. Pease and Amy Kaplan, 557–80. Durham, NC: Duke
University Press, 1993.

Pease, Donald. E. "Sublime Politics." In *The American Sublime*, edited by Mary
Arensberg, 21–50. Albany: State University of New York Press, 1986.

Pease, Donald, and Rob Wilson. "A Conversation with Ōe Kenzaburō." *boundary
2*, vol. 20, no. 2 (Summer 1993): 1–23.

Pendell, Dale. *Walking with Nobby: Conversations with Norman O. Brown*. San
Francisco: Mercury House, 2008.

Pentland, William. "Vast Freshwater Reserves Discovered under Ocean Floor,
Scientists Say." *Forbes*, December 5, 2013. http://www.forbes.com/sites
/williampentland/2013/12/05/epic-freshwater-reserves-discovered-under
-ocean-floor.

Perlman, Michael. *Imaginal Memory and the Place of Hiroshima*. Albany: State
University of New York Press, 1988.

Peter, Joachim. "Chuukese Travellers and the Idea of Horizon." *Asia Pacific View-
points* 41 (2002): 253–67.

Phelan, Joseph. "Italy's Plan to Save Venice from Sinking." BBC News, September 27, 2022. https://www.bbc.com/future/article/20220927-italys-plan-to
-save-venice-from-sinking.

Philip, Nourbese M. *Zong!* Middletown, CT: Wesleyan University Press, 2011.

Pompallier, Jean-Baptiste François. *Early History of the Catholic Church in Oceania*. Auckland: H. Brent, 1888.

Pope Francis I. *Laudato Si: On Care for Our Common Home* (encyclical letter),
June 18, 2015. https://www.vatican.va/content/francesco/en/encyclicals
/documents/papa-francesco_20150524_enciclica-laudato-si.html.

Probyn, Elspeth. *Eating the Ocean*. Durham NC: Duke University Press, 2016

Pukui, Mary Kawena, and Alfons Korn, eds. *The Echo of Our Song: Chants and
Poems of the Hawaiians*. Honolulu: University of Hawai'i Press, 1979.

Pule, John. *The Shark That Ate the Sun*. Auckland: Penguin, 1992.

Purdy, Jebediah. *After Nature: A Politics for the Anthropocene*. Cambridge, MA:
Harvard University Press, 2018.

Radhakrishnan, Radha. *History, the Human, and the World Between*. Durham,
NC: Duke University Press, 2008.

Rankine, Claudia. *Citizen: An American Lyric*. Minneapolis: Graywolf, 2014.

Reddy, Sujani, and Manu Vimalssery, eds. *The Sun Never Sets: South Asian Migrants in an Age of US Power*. New York: New York University Press, 2013.

Rice, Doyle. "Coral Reefs at Severe Risk as World's Oceans Become More Acidic."
USA Today, March 14, 2018. https://www.usatoday.com/story/tech
/science/2018/03/14/coral-reefs-severe-risk-worlds-oceans-become-more
-acidic/425017002.

Roberts, Brian Russell. *Borderwaters: Amid the Archipelagic States of America*.
Durham, NC: Duke University Press, 2021.

Roberts, Brian Russell, and Michelle Ann Stephens, eds. *Archipelagic American
Studies*. Durham, NC: Duke University Press, 2017.

Robin, Libby. "Three Galleries of the Anthropocene." *Anthropocene Review* 1, no. 3 (2014): 207–24. https://doi.org/10.1177/2053019614550533.

Robinson, Kim Stanley. "California: The Planet of the Future; The *Boom* Interview." *Boom* 3, no. 4 (Winter 2013): 3–11. https://doi.org/10.1525/boom .2013.3.4.3.

Robinson, Kim Stanley. "Empty Half the Earth of Its Humans. It's the Only Way to Save the Planet." *Guardian*, May 20, 2018. https://www.theguardian.com /cities/2018/mar/20/save-the-planet-half-earth-kim-stanley-robinson.

Ronda, Margaret. *Remainders: American Poetry at Nature's End*. Stanford, CA: Stanford University Press, 2018.

Roy, Ananya. "Conclusion: Postcolonial Urbanism: Speed, Hysteria, Mass Dreams." In *Worlding Cities: Asian Experiments and the Art of Being Global*, edited by Ananya Roy and Aihwa Ong, 307–35. Chichester, UK: Wiley-Blackwell, 2011.

Roy, Ananya, and Aihwa Ong, eds. *Worlding Cities: Asian Experiments and the Art of Being Global*. Chichester, UK: Wiley-Blackwell, 2011.

Roy, Arundhati. "Edward Snowden Meets Arundhati Roy and John Cusack." *Guardian*, November 28, 2015. http://www.theguardian.com/lifeandstyle/2015/nov /28/conversation-edward-snowden-arundhati-roy-john-cusack-interview.

Sacks, Sam. "In the Watery Part of the World." *Wall Street Journal*, March 19, 2012. https://www.wsj.com/articles/SB10001424052970204603004577271283247107866.

Said, Edward. *Orientalism*. New York: Vintage, 1994.

Saijo, Albert. OUTSPEAKS: *A Rhapsody*. Honolulu: Bamboo Ridge, 1997.

Sakai, Naoki. "Return to the West / Return to the East: Watsuji Tetsuro's Anthropology and Discussions of Authenticity." In *Japan in the World*, edited by Masao Miyoshi and Harry D. Harootunian, 237–70. Durham, NC: Duke University Press, 1993.

Sakaki, Nanao. *Break the Mirror*. Nobleboro, ME: Blackberry, 1996.

Sakaki, Nanao. *How to Live on Planet Earth: Collected Poems*. Nobleboro, ME: Blackberry, 2013.

Saldivar, José David. *Trans-Americanicity: Subaltern Modernities, Global Coloniality, and the Cultures of Greater Mexico*. Durham, NC: Duke University Press, 2012.

Sandhu, Sukhdev. "Allan Sekula: Filming the Forgotten Resistance at Sea." *Guardian*, April 20, 2012. http://www.guardian.co.uk/film/2012/apr/20/allan -sekula-resistance-at-sea.

Santayana, George. *Persons and Places: The Background of My Life*. New York: Scribner, 1944.

Santos Perez, Craig. *from Unincorporated Territory [hacha]*. Honolulu: Tinfish, 2008.

Santos Perez, Craig. *from Unincorporated Territory [saina]*. Richmond, CA: Omindawn, 2010.

Santos Perez, Craig. *from Unincorporated Territory [guma]*. Richmond, CA: Omindawn, 2014.

Sassen, Saskia. *Expulsions: Brutality and Complexity in the Global Economy.* Cambridge, MA: Harvard University Press, 2014.

Sassen, Saskia. *Guests and Aliens.* New York: New Press, 1999.

Schalansky, Judith. *Atlas of Remote Islands: Fifty Islands I Have Never Set Foot On and Never Will.* Translated by Christine Lo. New York: Penguin, 2010.

Schmitt, Carl. *Land and Sea: A World Historical Meditation.* Translated by Samuel Garrett Zeitlen. Candor, NY: Telos, 2015.

Schmitt, Carl. "The Monroe Doctrine as a Precedent for a *Grossraum* Principle." In *Writings on War*, translated by Timothy Nunan, 83–90. Cambridge: Polity, 2011.

Schmitt, Carl. *The Nomos of the Earth in the International Law of the Jus Publicum Europaeum.* Translated by G. L. Ulmen. Candor, NY: Telos, [1950] 2003.

Schwenger, Peter, and John Wittier Treat. "America's Hiroshima, Hiroshima's America." *boundary 2*, vol. 21, no. 1 (1994): 233–53.

Shanken, Andrew M. *Into the Void Pacific: Building the 1939 San Francisco World's Fair.* Berkeley: University of California Press, 2014.

Shell, Marc. *Islandology: Geography, Rhetoric, Politics.* Stanford, CA: Stanford University Press, 2014.

Shepard, Benjamin, and Greg Smithsimon. *The Beach beneath the Streets: Contesting New York City's Public Spaces.* Albany: State University of New York Press, 2011.

Shiels, Maggie. "Boat Made of Trash [the Plastiki] Prepares to Set Sail." BBC News, March 3, 2010.

Shigematsu, Setsu, and Keith L. Camacho, eds. *Militarized Currents: Toward a Decolonized Future in Asia and the Pacific.* Minneapolis: University of Minnesota Press, 2010.

Shu, Yuan, and Donald E. Pease, eds. *American Studies as Transnational Practice: Turning towards the Transpacific.* Dartmouth, NH: University Press of New England, 2016.

Silva, Noenoe K. *Aloha Betrayed: Native Hawaiian Resistance to American Colonialism.* Durham, NC: Duke University Press, 2004.

Simons, Paul. "Weather Eye: Disappearing Fogs of the Pacific Coast." *London Times*, February 25, 2010.

Siu, Helen F. "Retuning a Provincialized Middle Class in Asia's Urban Postmodern: The Case of Hong Kong." In *Worlding Cities: Asian Experiments and the Art of Being Global*, edited by Ananya Roy and Aihwa Ong, 127–59. Chichester, UK: Wiley-Blackwell, 2011.

Sloterdijk, Peter. *In the World Interior of Capital: Towards a Philosophical Theory of Globalization.* Translated by Wieland Hobanis. Cambridge: Polity, 2013.

Sloterdijk, Peter. *Neither Sun nor Death.* Translated by Steve Corcoran. New York: Semiotext(e), 2011.

Sloterdijk, Peter. *You Must Change Your Life*. Translated by Wieland Hoban. Cambridge: Polity, 2014.

Snyder, Gary. "Coming into the Watershed." In *A Place in Space: Ethics, Aesthetics, and Watersheds*, 219–35. Washington, DC: Counterpoint, 1995.

Snyder, Gary. *Earth House Hold: Technical Notes and Queries to Fellow Dharma Revolutionaries*. New York: New Directions, 1969.

Snyder, Gary. *Mountains and Rivers without End*. Washington, DC: Counterpoint, 1997.

Snyder, Gary. *The Old Ways: Six Essays*. San Francisco: City Lights, 1977.

Snyder, Gary. *A Place in Space: Ethics, Aesthetics, and Watersheds*. Berkeley, CA: Counterpoint, 1995.

Snyder, Gary. *The Practice of the Wild*. Berkeley, CA: Counterpoint, 2010.

Snyder, Gary. *Regarding Wave*. New York: New Directions, 1970.

Snyder, Gary. *Turtle Island*. New York: New Directions, 1974.

Solnit, Rebecca. *Hollow City: The Siege of San Francisco and the Crisis of American Urbanism*. London: Verso, 2000.

Solnit, Rebecca. *Infinite City: A San Francisco Atlas*. Berkeley: University of California Press, 2010.

Solnit, Rebecca. *Wanderlust: A History of Walking*. New York: Penguin, 2001.

Somerville, Alice Te Punga. "The Great Pacific Garbage Heap as Metaphor: The (American) Pacific You Can't See." In *Archipelagic American Studies*, edited by Brian Russell Roberts and Michelle Ann Stephens, 320–38. Durham, NC: Duke University Press, 2017.

Song, Min Hyoung. *Climate Lyricism*. Durham, NC: Duke University Press, 2022.

Spahr, Juliana. *This Connection of Everyone with Lungs*. Berkeley: University of California Press, 2005.

Spahr, Juliana. *Well Then There Now*. Boston: Godine, 2011.

Spahr, Juliana, and David Buuck. *An Army of Lovers*. San Francisco: City Lights, 2013.

Spicer, Jack. *My Vocabulary Did This to Me: The Collected Poetry of Jack Spicer*. Edited by Peter Gizzi and Keven Killian. Middletown, CT: Wesleyan University Press, 2008.

Spillers, Hortense. "Mama's Baby, Papa's Maybe: An American Grammar Book." *Diacritics* 17 (1987): 64–81.

Spivak, Gayatri. *Death of a Discipline*. New York: Columbia University Press, 2005.

Spivak, Gayatri. *Other Asias*. New York: Wiley, 2008.

Stamper, Gary. "At the Very Least Your Days of Eating Pacific Ocean Fish Are Over." *Collapsing into Consciousness* (blog), August 16, 2013. http://www.collapsingintoconsciousness.com/at-the-very-least-your-days-of-eating-pacific-ocean-fish-are-over.

Stein, Marc Eliot. "Diane di Prima's *Revolutionary Letters*." *Literary Kicks*, October 26, 2020. https://www.litkicks.com/RevolutionaryLetters.

Stein, Murray. *Jung's Map of the Soul: An Introduction*. Peru, IL: Open Court, 1998.

BIBLIOGRAPHY

Steinberg, Philip E. *The Social Construction of the Ocean.* Cambridge: Cambridge University Press, 2001.

Stevens, Wallace. *The Collected Poems of Wallace Stevens.* New York: Alfred A. Knopf, 1971.

Stevens, Wallace. *Opus Posthumous: Poems, Plays, Prose.* New York: Vintage, 2011.

Stevenson, Jim. "China Makes Debut in RIMPAC Naval Drills." *Voice of America,* July 4, 2014. http://www.voanews.com/content/china-makes-debut-in -international-naval-exercises/1950699.html.

Sullivan, Robert. *Star Waka: Poems.* Auckland: Auckland University Press, 1999.

Suzuki, Erin. "Transpacific." In *Routledge Companion to Asia American and Pacific Island Literature,* edited by Rachel C. Lee, 352–64. London: Routledge, 2014.

Swanson, Ana. "What America's Immigrants Looked Like When They Arrived on Ellis Island." *Washington Post,* October 24, 2015.

Tally, Robert, and Melody Yunzi Lee, eds. *Remapping the Homeland: Affective Geographies and Cultures of the Chinese Diaspora.* London: Palgrave Macmillan, 2022.

Taylor, Bayard. *Eldorado: Adventures in the Path of Empire.* New York: G. P. Putnam, 1850.

Teaiwa, Teresia Kieua. "bikinis and other s/pacific n/oceans." *Contemporary Pacific* 6 (1994): 87–109.

Teaiwa, Teresia Kieua. "Reading Imperialism in the Pacific: The Prose of Joseph Veramu and the Poetry of Sia Figiel." In *Anglo-American Imperialism and the Pacific: Discourses of Encounter,* ed. Michelle Keown, Andrew Taylor, and Mandy Treagus, 49–68. London: Routledge, 2019.

Teaiwa, Teresia Kieua. *Searching for Nei Nim'anoa.* Suva, Fiji: Mana, 1995.

Teaiwa, Teresia Kieua. *Sweat and Salt Water: Selected Works.* Edited by Katerina Teaiwa, April K. Henderson, and Terence Wesley-Smith. Honolulu: University of Hawai'i Press, 2021.

Teaiwa, Teresia Kieua, and Vilsoni Hereniko. *Last Virgin in Paradise: A Serious Comedy.* Suva, Fiji: Mana, 1993.

Thorne, Christian. "The Sea Is Not a Place, or Putting the World Back into World Literature." *boundary 2,* vol. 40, no. 2 (2013): 53–79.

Trend, David. *Worlding: Identity, Media, and Imagination in a Digital Age.* Boulder, CO: Paradigm, 2013.

Tsing, Anna Lowenhaupt. *The Mushroom at the End of the World: On the Possibility of Life in Capitalist Ruins.* Princeton, NJ: Princeton University Press, 2015.

Twain, Mark. *Following the Equator: A Journey around the World.* New York: P. F. Collier and Sons, 1899.

Urbain, Jean-Didier. *At the Beach.* Translated by Catherine Porter. Minneapolis: University of Minnesota Press, 2003.

Urbina, Ian. "'Sea Slaves': Forced Labor for Cheap Fish." *New York Times,* July 27, 2015. http://www.nytimes.com/2015/07/27/world/outlaw-ocean-thailand -fishing-sea-slaves-pets.html.

Vidal, Gore. "Requiem for the American Empire." *Nation*, January 11, 1986, 15–19.

Walker, Richard. *The Country in the City: The Greening of the San Francisco Bay Area*. Seattle: University of Washington Press, 2007.

Walters, Joanna. "Author Junot Diaz Called Unpatriotic as Dominican Republic Strips Him of Award." *Guardian*, October 25, 2015.

Wang, Chih-ming. "'Dreams of Colliding Worlds': Worlding Multiculturalism in Lawrence Chua's *Gold by the Inch*." In *Worlding Multiculturalisms: The Politics of Inter-Asian Dwelling*, edited by Daniel P. S. Goh, 17–34. London: Routledge, 2015.

Wang, Chih-ming. *Transpacific Articulations: Student Migration and the Making of Asian America*. Honolulu: University of Hawai'i Press, 2013.

Wang, Hui. *China's New Order: Society, Politics, and Economy in Transition*. Edited by Theodore Huters. Cambridge, MA: Harvard University Press, 2003.

Wang, June, Tim Oakes, and Yang Yang, eds. *Making Cultural Cities in Asia: Mobility, Assemblage, and Aspirational Urbanism*. Routledge: New York, 2018.

Wark, McKenzie. *The Beach beneath the Street: The Everyday Life and Glorious Times of the Situationist International*. New York: Verso, 2011.

Wark, McKenzie. *Molecular Red: Theory for the Anthropocene*. London: Verso, 2015.

Warner, Marina. "Here Be Monsters." *New York Review of Books*, December 19, 2013. http://www.nybooks.com/articles/archives/2013/dec/19/here-be -monsters/.

Watts, Jonathan. "Eight Months On, Is the World's Most Drastic Plastic Bag Ban Working?" *Guardian*, April 25, 2018. https://www.theguardian.com/world /2018/apr/25/nairobi-clean-up-highs-lows-kenyas-plastic-bag-ban.

Weart, Spencer R. *Nuclear Fear: A History of Images*. Cambridge, MA: Harvard University Press, 1988.

Weiss, Jeff. "Driving the Beat Road" (interview with Lawrence Ferlinghetti). *Washington Post*, June 30, 2017. https://www.washingtonpost.com/graphics /2017/lifestyle/the-beat-generation.

Wellman, Elizabeth Iam, and Loren B. Landau. "South Africa's Tough Lessons on Migrant Policy." *Foreign Policy*, October 13, 2015. https://foreignpolicy .com/2015/10/13/south-africas-tough-lessons-on-migrant-policy.

Wendt, Albert, Reina Whaitiri, and Robert Sullivan, eds. *Whetu Moana: Contemporary Polynesian Poems in English*. Honolulu: University of Hawai'i Press, 2003.

Wilkinson, Tracy, Shashank Bengali, and Brian Bennet. "Trump's New Foreign Policy Touts a Free and Open Indo-Pacific." *Los Angeles Times*, November 7, 2017. http://www.latimes.com/nation/la-fg-trump-indo-pacific -20171108-story.html.

Wilson, Rob. *American Sublime: The Genealogy of a Poetic Genre*. Madison: University of Wisconsin Press, 1991.

Wilson, Rob. *Be Always Converting, Be Always Converted: An American Poetics*. Cambridge, MA: Harvard University Press, 2009.

Wilson, Rob. "Bob Dylan in China, America in Bob Dylan: Visions of Social Beatitude and Critique." *boundary 2*, vol. 41, no. 3 (2014): 159–78.

Wilson, Rob. "Imagining Asia-Pacific Today: Forgetting Colonialism in the Magical Free Markets of the American Pacific." In *Learning Places: The Afterlives of Area Studies*, edited by Masao Miyoshi and Harry D. Harootunian, 231–60. Durham, NC: Duke University Press, 2002.

Wilson, Rob. "Milton Murayama's Working Class Diaspora across the Japanese/Hawaiian Pacific." *Postcolonial Studies* 11 (2008): 475–79.

Wilson, Rob. "Pacific Postmodern: From the Sublime to the Devious." *boundary 2*, vol. 28, no. 1 (2001): 121–51.

Wilson, Rob. *Pacific Postmodern: From the Sublime to the Devious, Writing the Local/Experimental Pacific in Hawai'i*. Honolulu: Tinfish, 2000.

Wilson, Rob. "Postcolonial Pacific Poetries: Becoming Oceania." In *The Cambridge Companion to Postcolonial Poetry*, edited by Jahan Ramazani. 58–71. Cambridge: Cambridge University Press, 2017.

Wilson, Rob. "Postmodern as Post-nuclear: Landscape as Nuclear Grid." In *Ethics/Aesthetics: Postmodern Positions*, ed. Robert Merrill 169–92. Washington DC: Maisonneuve, 1988.

Wilson, Rob. "Postmodern as Post-nuclear: Landscape as Nuclear Grid." *Prospects* 1 (1989): 407–39.

Wilson, Rob. "The Postmodern Sublime: Local Definitions, Global Deformations of the US National Imaginary." *Amerikastudien* 37 (1992): 613–45.

Wilson, Rob. "Reframing Global/Local Poetics in the Post-imperial Pacific: Meditations on 'Displacement,' Indigeneity, and the Misrecognitions of US Area Studies." In *World Writing: Poetics, Ethics, Globalization*, edited by Mary Gallagher, 224–45. Toronto: University of Toronto Press, 2009.

Wilson, Rob. *Reimagining the American Pacific: From "South Pacific" to Bamboo Ridge and Beyond*. Durham, NC: Duke University Press, 2000.

Wilson, Rob. "Review of Richard Hamasaki's *From the Spider Bone Diaries: Poems and Songs*." *Contemporary Pacific* 14, no. 2 (2002): 511–13.

Wilson, Rob. "Seven Tourist Sonnets." In *Jack London Is Dead: Contemporary Euro-American Poetry of Hawai'i (and Some Stories)*, edited by Susan Schultz. Kaneohe, HI: Tinfish, 2013.

Wilson, Rob. "*Snowpiercer* as Anthropoetics: Killer Capitalism, the Anthropocene, Korean-Global Film." *boundary 2*, vol. 46, no. 3 (2019): 199–218.

Wilson, Rob. "Spectral City: San Francisco as Pacific Rim City and Countercultural Contado." *Inter-Asia Cultural Studies* 9, no. 4 (2008): 583–97.

Wilson, Rob. "Sublime Patriot." *Polygraph* 5 (1992): 67–77.

Wilson, Rob. "Transfiguration as a World-Making Practice: From Norman O. Brown to Bob Dylan." *boundary 2*, vol. 49, no. 3 (2022): 99–116.

Wilson, Rob. "'World Gone Wrong': Thomas Friedman's *World Gone Flat* and Pascale Casanova's *World Republic of Letters* against the Multitudes of Oceania." *Concentrics* 33 (2007): 177–94.

Wilson, Rob. "Worlding as Future Tactic." In *The Worlding Project: Doing Cultural Studies in the Era of Globalization*, edited by Rob Wilson and Christopher Leigh Connery, 209–23. Berkeley, CA: North Atlantic and New Pacific, 2007.

Wilson, Rob, and Christopher Leigh Connery, eds. *The Worlding Project: Doing Cultural Studies in the Era of Globalization*. Berkeley, CA: North Atlantic and New Pacific, 2007.

Wilson, Rob, and Arif Dirlik, eds. *Asia/Pacific as Space of Cultural Production*. Durham, NC: Duke University Press, 1995.

Winduo, Steven Edmund. "Chief of Oceania." *Contemporary Pacific* 22 (2010): 114–15.

Wolfe, Alan. *Suicidal Narrative in Modern Japan: The Case of Dazai Osamu*. Princeton, NJ: Princeton University Press, 1990.

Woo-Cumings, Meredith. "Market Dependency in U.S.-East Asian Relations." In *What Is in a Rim? Critical Perspectives on the Pacific Region Idea*, edited by Arif Dirlik, 163–86. Lanham, MD: Rowman and Littlefield, 1998.

Wood, Houston. "Cultural Studies for Oceania." *Contemporary Pacific* 15 (2003): 340–74.

Wright, Richard. *Haiku: The Other World*. Edited by Yoshinobu Hakutani and Robert Tener. New York: Arcade, 1998.

Yamashita, Karen Tei. *Anime Wong: Fictions of Performance*. Minneapolis: Coffee House, 2014.

Yamashita, Karen Tei. *I Hotel*. Minneapolis: Coffee House, 2010.

Yamashita, Karen Tei. *Letters to Memory*. Minneapolis: Coffee House, 2017.

Yao, Steven. "Oceanic Etymologies: Shanghai and the Transpacific Routes of Global Modernity." *Verge* 3 (2017): 77–106.

Yeo, George. "For the Rest of Asia, America Might Be a Friend, but China Cannot Be an Enemy." *China/US Focus*, August 26, 2014. https://www.chinaus focus.com/foreign-policy/for-the-rest-of-asia-america-might-be-a-friend-but-china-cannot-be-an-enemy.

Žižek, Slavoj. "Eastern Europe's Republics of Gilead." *New Left Review* 183 (1990): 53–56.

Žižek, Slavoj. "Have Michael Hardt and Antonio Negri Rewritten the *Communist Manifesto* for the Twenty-First Century?" In Karl Marx, *The Communist Manifesto: Norton Critical Edition*, ed. Frederic L. Bender, 225–34. New York: W. W. Norton, 2013.

Žižek, Slavoj. *Looking Awry: An Introduction to Jacques Lacan through Popular Culture*. Cambridge, MA: MIT Press, 1991.

Žižek, Slavoj. *The Sublime Object of Ideology*. London: Verso, 1989.

INDEX

Abbas, Ackbar, 26
"Abomunist" (Kaufman), 114–15
Adams, John Luther, 43
Afghanistan, 128, 158
Africa, 3, 18, 58
After Empire (Gilroy), 100
Akira Kurosawa's Dreams (Kurosawa), 139–40
Alfaisal, Haifa Saud, 52
Aliens of the Deep (documentary), 41
Allen, Helena, 156
All I Asking for Is My Body (Murayama), 91
Alon, Shir, 61
"Always Keeping It Real" (Teaiwa), 156–60
"America" (Ginsberg), 115
"Amnesia" (Teaiwa), 82, 156–57
"Animal Rhapsode" (Saijo), 85
Anime Wong (Yamashita), 82
Anthropocene: Berlin, 7–11; capitalism and, 26; ecology of, 56; era, 130; flourishing in, 163; Haraway on, 108; history of, 4; Netherlands in, 23–24; nuclear, 177n3; reminders of, 1–2; rise of, 54; Robinson on, 169; threats of, 21–24; for United States, 55
"Any Fool Can Get into an Ocean" (Spicer), 23
Anzaldúa, Gloria, 102–3
Apio, Alani, 151

Appadurai, Arjun, 74
Apple, 5, 117
Arac, Jonathan, 72
Arendt, Hannah, 61, 62, 101
Army of Lovers, An (Buuck and Spahr), 114
Asia as Method (Chen), 63
Asia Pacific (Asia/Pacific): deworlding in, 27, 55; forces, 90; Global China and, 54; hub, 5–6; Oceania and, 60–67; Pacific Rim and, 35, 82; scholarship on, 65, 78–80, 83; worlding, 51–53, 71–80, 177n2
Asia/Pacific as Space of Cultural Production (collection), 79, 91
Asia-Pacific Economic Cooperation, 140, 149, 192n14
"Asia Pacific Studies in an Age of Global Modernity" (Dirlik), 78–79
Asia sublates Pacific, 80–88
Assange, Julian, 102
Atlantic Ocean, 2–4, 9–10, 16–20, 22–23, 26, 32–34, 43–45
Atlas of Remote Islands (Schalansky), 22
Atop an Underwood (film), 46
Auckland. *See* New Zealand
Australia, 66–67, 71–72, 80–81, 89, 142, 187n33

Baez, Joan, 117
Balaz, Joseph Puna, 27, 80–81, 83, 89, 142–43,
    193n5
Ballard, J. G., 128–32, 139
Balog, Jason, 66
Barad, Karen, 27, 60
Barnett, David, 47
Baudelaire, Charles, 12
Bauman, Zygmunt, 96
Beach beneath the Street, The (Wark), 8–9, 12
Be Always Converting, Be Always Converted
    (study), 39, 74, 83
Beat culture: in culture, 91, 94–95, 99; ethics
    of, 88, 105; Kaufman for, 49, 114–16; legacy
    of, 12, 18–19, 25, 45–46; in San Francisco,
    120, 122–23
Benjamin, Walter, 102
Bernes, Jasper, 7, 114
Bernstein, Charles, 37
Biden, Joe, 95, 103
Big Sur (Kerouac), 47
Bikini Atoll, 14, 32–34, 114–15
Black Rain (Ibuse), 138–39
Black Wings (Raporgan), 84
Blade Runner (film), 6–7
Blowback (Johnson), 149
Blue Crush (film), 15
blue ecopoetics, 24–25, 71–78, 80–88, 90–91
Blue Hawaii (film), 15
blue humanities, 5, 11, 13, 15–16, 33, 74, 167n1
blue transfiguration, 78–80
Bomboki, Oilei, 81
Bong Joon-Ho, 55, 93, 150
Borderlands / La Frontera (Anzaldúa), 102–3
border water, 3–4, 71–72, 79–80, 165–66
Borderwaters (Roberts), 162
Bourne, Randolph, 95
Brathwaite, Edward Kamau, 40, 72, 143
Brautigan, Richard, 114–15, 119–20
Brechin, Gray, 112, 118
Brislin, Tom, 10–11
Brown, Norman O., 11, 76, 163
Buell, Lawrence, 90
Burke, Edmund, 20
Burroughs, William, 14, 45, 47
Bush, George W., 145
Buuck, David, 114

California: coastal, 2–5, 8, 10–14, 51–52,
    124–25; culture of, 100–106; geography of,
    34; Hawai'i and, 42, 72; Japan and, 31; to
    United States, 46. *See also specific locations*

Call Me Ishmael (Olson), 35
capitalism: Anthropocene and, 26; axiomatic
    of, 194n34; climate change and, 178n22;
    communism and, 106–7; deworlding and,
    51–60; ethos of, 193n12; expansion of,
    9; global, 4, 16, 61, 100–101, 143–44; in
    globalization, 59–60; late, 54; modern, 19;
    Occupy Movement against, 12; philosophy
    of, 19; planetary, 53, 57; transpacific, 80;
    urban space in, 7
Caribbean culture, 39–40, 53
Carson, Rachel, 1–2, 48, 166
Casanova, Pascale, 21, 62
Catholic Church, 2, 74–76, 87, 90
Certeau, Michel de, 86
Cha, Theresa, 102
Chamisso, Adelbert von, 73
Channer, Colin, 43–44
"Chants Democratic" (Whitman), 155
Chasing Coral (documentary), 66
Chasing Ice (documentary), 66
Cheah, Pheng, 52–53, 61–62, 178n11
Cheever, John, 40
Chen, Kuan-Hsing, 63, 64, 79
Ch'ien, Evelyn, 141
Child, Ben, 174n30
Chile, 142, 192n14
China: in Cold War, 162; Global, 5, 80, 148;
    in globalization, 55–59, 173n8, 187n28;
    hegemony of, 14, 36; Japan and, 6–7, 25,
    35, 42, 103, 132–33, 145; leadership in, 18,
    31–32, 35; mainland, 83–84; Russia and, 80,
    128; Singapore and, 43, 52, 64; South Korea
    and, 73–74; Taiwan and, 93–94, 107; United
    States and, 22, 33–34, 95, 104–5, 173n4,
    174n30, 176n43
Chong, Terence, 65
Chou, Shiuhhuah Serena, 52, 124
Christianity, 74–75
Chronicles (Dylan), 144
Chua, Beng-Huat, 63
Chuah, Lawrence, 60
Citizen (Rankine), 23
City Lights Books/Press, 1–2, 86, 114, 115–17
Cleveland, Grover, 153–54
Cliff, Michelle, 102
Clifford, James, 89–90, 98
climate change, 2–3, 7, 17–18, 36–37, 72,
    90–93, 178n22
Clinton, Hillary, 36
Cloud Atlas (Mitchell), 14–15, 37
Clover, Joshua, 1, 114

coastal California, 2–5, 8, 10–14, 51–52, 124–25
Cohen, Jeffrey, 17
Cold Sea (Raporgan), 84
Cold War. *See specific topics*
Coleman, Ornette, 163
Coleridge, Samuel Taylor, 46
colonialism, 4, 19, 40, 74–75, 80
Columbus, Christopher, 17–18
"Coming into the Watershed" (Snyder), 88, 120–21
communism, 80, 93, 95–96, 106–7
Communist Manifesto, The (Marx), 95–96
Connecticut, 163–66
"Connect/I/cut" (Deleuze), 163–65
Connery, Christopher Leigh, 5, 33, 59, 81, 173n4
Conrad, Joseph, 20, 40
conversion, 11, 66, 71–78, 88, 111–12, 120, 164
"cool beatitude" (Kaufman), 116
Coppola, Sofia, 148
coral reefs, 15–19, 66–67, 180n85
Country and the City, The (Williams), 118–19
Crane, Stephen, 40
Crazy Rich Asians (film), 150
Creation of the World or Globalization, The (Nancy), 52–53
"Crisis Poem #2" (Teaiwa), 157
Cruise, Tom, 148

"Da Last Squid" (Balaz), 89
Damrosch, David, 62
Dana, Richard Henry, 47
Darwin, Charles, 37
Davis, Miles, 163
Deep Passion (Raporgan), 84
Defoe, Daniel, 13
Deleuze, Gilles, 46–47, 157, 159–60, 164, 165–66, 194n34
DeLong, Ed, 36
Demick, Barbara, 16, 41
de-Orientalizing, 61
Derrida, Jacques, 61–62, 127
deworlding, 18, 27; capitalism and, 51–60; refugees in, 92–94; reworlding and, 26, 57, 60–67, 80–81, 100, 157
Dharma Bums (Kerouac), 104, 122
Díaz, Junot, 102
Dimock, Wai Chee, 39, 51–52
Dirlik, Arif, 65, 78–80, 91
Disney, Walt, 37
Dole, Sanford, 153–54

Dominican Republic, 102
Doo-hwan, Chun, 11–12
Dorchester, ss (ship), 45–46, 47
Double Trio (Mackey), 162–63
Duncan, Robert, 163
d'Urville, Dumont, 40, 75–76
Dylan, Bob, 12, 17, 26, 46, 113; on Beat culture, 94–95; fans of, 117; in globalization, 144; legacy of, 74, 163

Earle, Sylvia, 90–91
Early History of the Catholic Church in Oceania (Pompallier), 76
Earth House Hold (Snyder), 77, 84, 88, 90, 120
ecological alerts, 73–74
ecological belonging, 81
ecological consciousness, 111–15, 124
ecological prefigurations, 84
ecological solidarity, 89–90
ecology, 3, 7–13, 16, 20, 23; of Anthropocene, 56; to Balaz, 89; of climate change, 72; of coral reefs, 66–67; Guattari on, 182n23; of Oceania, 36–38; reworlding, 160; Snyder on, 76–77, 84–85; of Taiwan, 25
Economic Times, 6
ecopoetics, 14–15, 26, 32–33, 42–43, 120; blue, 24–25, 71–78, 80–88, 90–91; of transpacific solidarity, 88–90
Ehlers, Otto E., 21
Eisen-Martin, Tongo, 25–26, 108
Eldorado (Taylor), 116
Eliot, George, 20
El Niño, 2–3, 90–91
Emerson, Nathaniel, 155
Emerson, Ralph Waldo, 76, 94, 153
Empire (Hardt and Negri), 96, 98
Empire of the Sun (Ballard), 128–34, 139
"Empty Half the Earth of Its Humans" (Robinson), 141
"Epistemic Reading and the Worlding of Postcolonialism" (Alfaisal), 52
Europe, 3, 5, 9–10, 87; Americas and, 92–100; Australia and, 80; concepts of, 78; de-Europeanization, 13; Hawai'i and, 19; history of, 22, 177n2; refugees in, 102; theories from, 52

Facebook, 121
"Farewell Angelina" (Dylan), 117
Ferlinghetti, Lawrence, 1–4, 115–17, 122–23
Fifth Book of Peace, The (Kingston), 123
Figure of the Migrant, The (Nail), 92–93

Fish Story (Sekula), 40
"Flamenco Sketches," 163
Forgotten Space, The (Sekula), 40
Foucault, Michel, 80, 99
France, 2, 8–9, 96, 102
Francis I (pope), 18–19
Franklin, Benjamin, 60–61
Freewheeling Bob Dylan, The, 26
Freud, Sigmund, 76
from Unincorporated Territory (Santos
  Perez), 87, 89
Frost, Robert, 37
Fussell, Paul, 136–37

Galbraith, John Kenneth, 136–37
Garcia, Edgar, 5
Gauguin, Paul, 21
Germany, 3, 13, 23, 34–35; activism in, 19;
  Berlin Anthropocene, 7–11; culture of,
  9–10, 20–21, 190n5; France and, 2, 96, 102;
  history of, 21–22, 175n34; Italy and, 96–97;
  Poland and, 93; in World War II, 45–46
Ghosh, Amitav, 56, 60
Ghost in the Shell (film), 6–7
"Giant Bumptious Litter, A" (Haraway), 51
"Gift Outright, The" (Frost), 37
Gillman, Susan, 167n6
Gilroy, Paul, 100, 104
Ginsberg, Allen, 11, 14, 23–24, 112; Kerouac
  and, 45–46, 104, 123, 176n55; legacy of,
  114–15, 120
Giroux, Robert, 47
Glass Palace, The (Ghosh), 56
Glissant, Édouard, 19–20, 40, 53, 59,
  177nn9–10
global cities: in California, 12–13; dominance
  of, 9; ecocide in, 18–19; fake, 56; Great
  Pacific Garbage Patch and, 113; leadership
  in, 101; Mirzoeff on, 169n34; oceans and,
  22–23; pavement in, 1–2, 26; in planetary
  capitalism, 53; reworlding, 86–87; waste
  in, 17–18
global flows, 92–100
globalization: Africa in, 18; China in, 55–59,
  173n8, 187n28; deworlding in, 52–53; diaspora discourse in, 160; discourse, 75; Dylan
  in, 144; food resources in, 4–5; global
  capitalism, 4, 16, 61, 100–101, 141–44;
  global citizenship, 95–96; hegemonic,
  3; hyper-globalization, 92–93; localization and, 3–4, 11–12; monoculture from,
  184n51; neoliberalism and, 185n70; Oceania

in, 33–34, 41–43, 51–60; Pacific Rim in,
  174n30; for refugees, 99–100; of Silk Road,
  31–32; Taiwan in, 33; United States in,
  97–98, 126–32; wet, 42; worlding in, 148;
  after World War II, 103–4
Global South, 26, 101
Gluck, Carol, 57
Goethe, Johann Wolfgang von, 20–21, 60, 62
Goh, Daniel P. S., 64
Gold by the Inch (Chuah), 60
Gonick, Sophie, 12–13
Google, 116, 121
Gramsci, Antonio, 89
Gray, Timothy, 120
Great Barrier Reef, 66–67
Great Derangement, The (Ghosh), 56
Great Ocean, 72–73
Great Pacific Garbage Patch, 4, 14, 16, 24,
  39–42, 73, 113
Greece, 96–97
Guam, 39, 43, 73–74, 85–87
Guattari, Félix, 160, 182n23, 194n34
Guests and Aliens (Sassen), 96
Gulf War, 127–28

Hadot, Pierre, 74–75
Haines, Chad, 63
Hall, Stuart, 89
Hamasaki, Richard, 91
Handel, George Frideric, 163
Haraway, Donna, 5, 27, 51, 53–54, 57–58, 108,
  168n16
"Hard Rain's A-Gonna Fall, A" (Dylan), 17, 26
Hardt, Michael, 96, 98
Harvey, David, 54–56, 96
Hau'ofa, Epeli, 19–20, 26–27; Brown and,
  76; colleagues of, 36; ecopoetics of, 77–78;
  field composition poems from, 86; in glo-
  balization discourse, 75; on Great Pacific
  Garbage Patch, 39–40; on Oceania, 65–66,
  75–77, 81–82, 89–90, 174n24; oceanic social
  theory from, 72; oceans to, 80; religion to,
  90; satire from, 74, 80–81, 83–84
Hawai'i: Australia and, 89; California and, 42,
  72; in culture, 142–43; culture of, 27, 159;
  Europe and, 19; Hawai'i Dub Machine, 85;
  history of, 6, 9–12, 14, 38–39; Japan and,
  14, 34–35; Lili'uokalani for, 27, 150–56;
  McDougall on, 38–39; Oceania and, 160;
  Philippines and, 106; religion in, 156; social
  justice in, 73; South Korea and, 25; Taiwan
  and, 98; Teaiwa and, 194n29; tourism

in, 10–11; United States and, 37, 150–56; University of Hawai'i, 142, 163–64, 183n45; volcanos in, 75–76

Hawaii's Story by Hawaii's Queen (Lili'uokalani), 150–56

Hayakawa, S. I., 106

Hayot, Eric, 52, 57–60

Heaven Is All Goodbyes (Eisen-Martin), 108

He Buke Mele Hawaii (song book), 150–51

Hegel, G. W. F., 20–21, 34, 61

Heidegger, Martin, 61, 179n28

"He Inoa Ahi no Ka-la-kaua" (Malo, D.), 155–56

Helm, George, 151

Helmreich, Stefan, 42

"He Mele Lāhui Hawai'i," 151

He Pule Hoolaa Alii (Kalākaua), 154–55

Herder, Johann Gottfried von, 153

Hereniko, Vilsoni, 79–80, 183n45

Hersey, John, 134, 139

Hillman, James, 129

Hiroshima, 2–4, 7, 8, 126–30, 134–40. *See also* Japan

Hiroshima (Hersey), 134, 139

Hiroshima (TV movie), 138–39

"Hiroshima" (Fussell), 136

Hiroshima Notes (Ōe), 128, 134–36, 140

Hitchcock, Alfred, 116

Ho, Vincent, 65

Hollow City (Solnit), 119

Holt, John Dominis, 151

Hooker, Thomas, 164

Hopkins, Gerard Manley, 4, 162, 166

Host, The (film), 41

Howl (Ginsberg), 11, 104, 112, 114

Hsu, Hua, 6

Huang, Hsinya, 84, 90, 124

Hull, Christopher, 138–39

Hume, Keri, 21

Hungry Shore, The (Ghosh), 56

Ibuse, Masuji, 130, 138

I Hotel (Yamashita), 82, 104–8, 124

Imaginal Memory and the Place of Hiroshima (Perlman), 136

"Imagining Asia Pacific" (collection), 83

immigration, 10, 98–99, 101, 104–6, 188n51

Imperial San Francisco (Brechin), 118

Inada, Lawson, 82–83

India, 6, 87–88, 120

"Indigenous Articulations" (Clifford), 89

inter-Asia, 52, 54–55, 63–64, 78, 145

In the South Seas (Stevenson), 20

"Invisible Pacific, The," 8

Iraq, 128, 158

Isle of Dogs (film), 6–7

Jameson, Fredric, 129–30, 141

Jan Ken Po, 85

Jankowski, Martin, 9

Japan: California and, 31; China and, 6–7, 25, 35, 42, 103, 132–33, 145; Fukushima disaster, 4; Germany and, 35; Guam and, 73–74, 86–87; Hawai'i and, 14, 34–35; history of, 126–32; India and, 120; nuclear sublime in, 132–40, 190n4, 191n6; Santa Cruz and, 34; South Korea and, 36; Taiwan and, 72, 84; United States and, 7, 34, 41, 123, 136–40, 192n14; in World War II, 2, 5–6, 22

Jarosinski, Eric, 168n25

Jobs, Steve, 113, 117

Johnson, Chalmers, 149

John Wesley Hardin, 94–95

John XXIII (pope), 76

Jung, Carl, 16

Kalākaua, David (king), 152–55

Kamaka'eha, Lydia, 152–53

Kamakau, Samuel, 155

Kamehameha V (king), 151

Kanai, Toshihiro, 134

Kant, Immanuel, 20

Kaplan, Robert D., 42

Kaufman, Bob, 49, 104, 114–17, 120, 123–24

"Ka Wiliwili Wai" (Lili'uokalani), 154

Kennedy, John F., 37, 134

Kerouac, Jack, 16–17, 44–49, 72, 104, 120, 122–24, 176n55

Killian, Tom, 119

Kim, Annette M., 101

Kim, Sung Kyung, 65

Kincaid, Jamaica, 102

Kingston, Earll, 123

Kingston, Maxine Hong, 104, 123–24

Kirksey, Eben, 53

Kisses in the Nederends (Hau'ofa), 81

Kneubuhl, Victoria, 151

Knott, Bill, 18–19

Krushchev, Nikita, 134

Kumilipo (Kalākaua), 38, 154–56

Kurosawa, Akira, 139–40

Lamming, George, 102

Lang, Fritz, 22

La Niña, 2–3

Lapham, Lewis, 20
Last Black Man in San Francisco, The (film), 108
Last Samurai, The (film), 148
Lawrence, D. H., 169n30
Ledyard, John, 164
Lee, Chang-rae, 14–15
Lee, Erika, 103, 107
Lee, Zack, 63
Legends and Myths of Hawaii (Kalākaua), 155
Leonard, Annie, 111, 113
Leopold, Aldo, 102
Lepawsky, Josh, 58
Les Fleurs du Mal, 12
Letzel, Jan, 140
Life of Pi (Martel), 14–15, 37
Liliʻuokalani (queen), 27, 146, 150–56
Lim, Alvin, 18
Li Po, 25
Loba (di Prima), 161–62
localization, 3–4, 11–14, 38–39, 57–59, 71–74, 96–98, 107–8
London, Jack, 40
Lonesome Traveler (Kerouac), 47
Longfellow, Henry Wadsworth, 153
Lost in Translation (film), 148
"Love Supreme, A" (Coleman), 163
Lynas, Mark, 15–16

MacArthur, Douglas, 55
Mackey, Nathaniel, 92–93, 162–63
Magellan, Ferdinand, 2, 17–18
Mahan, Alfred Thayer, 21, 42
Making of Asian America, The (Lee, E.), 103–4
Malo, David, 155–56
Malo, Juan, 87
"Mama's Baby, Papa's Maybe" (Spillers), 71
Mandel, Ernst, 130
Mao Zedong, 35, 106, 145–46
mappamundi maps, 16
maps, of Pacific Ocean, 20–23
Martel, Yann, 14–15
Marx, Karl, 60–61, 76, 95–96, 177n7
Marx, Leo, 118–19
Marxism, 12, 118
Mather, Cotton, 116
Mattis, James, 6
McDougall, Brandy Nalani, 14–15, 36, 38–39, 89
Melville, Herman, 1–2, 15, 19–20, 35–37, 40–41, 46, 173n5
Mentz, Steve, 14, 33–34, 73–74

Merchant of Venice (Shakespeare), 19
Merwin, William S., 143
Messiah, 163
Metropolis (film), 22
Mexico, 47, 94, 95, 98, 102–3
Michelet, Jules, 13
Michener, James, 40, 164
Middle Passage, 43–44
Middle Passages (Brathwaite), 72–73
migrants: to Australia, 187n33; migrancy, 103–8; migrant blockages, 92–100, 108; migrant frames, 100–103; Nail on, 87n34; in postcolonialism, 187n25; sea labor, 16; worlding and, 24–25
Miles, Josephine, 164
Mirzoeff, Nick, 9, 169n34
Mitchell, David, 14–15
Miyoshi, Masao, 13
Moby-Dick (Melville), 15, 20, 31–32, 41, 149, 152
Molecular Red (Wark), 55
Monroe Doctrine, 35, 151, 157
Monterey Bay, 32, 71–78
Moore, Charles, 20
Moore, Marianne, 17
Moretti, Franco, 62
Morris, Meaghan, 129–30
Mountains and Rivers without End (Snyder), 85
Mukerjee, Bharati, 102
Murayama, Milton, 91
Mushroom at the End of the World, The (Tsing), 56–57
"Myth of My Own Village, The" (Ōe), 126–27

Nagasaki, 2, 8, 20, 123, 126–27, 130–32, 191n6
Nail, Thomas, 87n34, 92–93, 99, 101
Nancy, Jean-Luc, 52–53, 56, 59, 177n7, 185n70
"Nani Nā Pua," 151
Negri, Antonio, 96, 98
Neither Sun nor Death (Sloterdijk), 17–18, 41–42
neoliberalism, 36, 96, 100–101, 185n70
neo–Silk Road, 54–55
Netherlands, 23–24
"New World Water," 44
New York City, 60–61, 97, 108. *See also* global cities
New Zealand, 72, 81
Nguyen, Viet, 100
Nixon, Richard, 35
Nixon, Rob, 36, 94, 102

Nolde, Emil, 21
"North Beach" (Snyder), 88–89, 121
North Korea, 55, 93, 128, 140
"Notes on Poetry as an Ecological Survival
    Tool" (Snyder), 77
nuclear Anthropocene, 177n3
nuclear disasters, 13–14
nuclear empires, 58
nuclear energy, 4, 136–37
nuclear refugees, 114–15
nuclear sublime, 132–40, 190n4, 191n6
nuclear technology, 115–17, 126–28, 130–32
nuclear testing, 1, 14, 32
nuclear weapons, 2, 7, 14, 33–34, 119–25,
    127–30

Obama, Barack, 36
Ocean (Mentz), 160
"Ocean Birth" (Sullivan), 86
ocean commons, 3, 5, 11, 34–35, 71–78, 90
Oceania: Asia Pacific and, 60–67; Asia sub-
    lates of, 80–88; blue ecopoetics of, 90–91;
    contemporary, 58; ecology of, 36–38;
    Global China and, 80; in globalization,
    33–34, 41–43, 51–60; Hau'ofa on, 65–66,
    75–77, 81–82, 89–90, 174n24; Hawai'i and,
    160; history of, 31–33; native peoples of, 26;
    ocean commons as, 71–78; Pacific Rim and,
    9, 142–43, 160; Pacific world of, 128; Papua
    New Guinea and, 149; planetary global, 157;
    poetics of, 39–41, 43–50, 86; postmodern,
    87; reworlding, 38–39; scholarship on, 3–4,
    88–90, 183n45; Teaiwa on, 156–60; United
    States in, 34–36; writing, 161–66
oceanic becoming. *See specific topics*
"Ocean in Us, The" (Hau'ofa), 81–82
Ōe, Kenzaburō, 126–28, 134–36, 138, 140,
    191n6
Olive, Jason, 165
Olson, Charles, 35, 61, 86
O'Neill, Eugene, 20
"1,000 Singapores" (Chua), 63
Ong, Aihwa, 53, 62–64
On Such a Full Sea (Lee, C.), 14–15
On the Road (Kerouac), 46, 122
Oppen, George, 116
Opukahai'a, Henry, 164
Orientalism, 74, 82, 148–49, 193n15
Orlowski, Jeff, 66
Oshii, Mamoru, 6–7
"Our Plastic Pollution Crisis Is Too Big for
    Recycling to Fix" (Leonard), 111

"Outer Banks, The" (Rukeyser), 86
"Out of the Cradle Endlessly Rocking" (Whit-
    man), 16–17
OUTSPEAKS (Saijo), 85, 90

Pacific Ocean: Atlantic Ocean and, 22, 26;
    Australia and, 71–72; in Berlin Anthropo-
    cene, 7–11; in Caribbean culture, 39–40;
    after Cold War, 79; to Europe, 9–10; as
    Great Ocean, 72–73; to Kerouac, 48–49;
    maps of, 20–23; ocean consciousness with,
    13–14; Robinson on, 169; scholarship on,
    1–7; to United States, 35–36, 43; urban
    space and, 23–27
Pacific Postmodern (Wilson), 141–42
Pacific Rim: Asia Pacific and, 35, 82; cities
    along, 7–8; climate change in, 2–3; com-
    munity, 120; cultures, 145–46; definitions
    of, 36; diversity of, 34; ecopoetics with,
    82–83; forces, 81; global capitalism in,
    143–44; Global China and, 148; in global-
    ization, 174n30; in Global South, 26; as
    Indo-Pacific, 95; modernism, 159; Oceania
    and, 9, 142–43, 160; San Francisco and,
    25; South Korea and, 125, 143–46; studies,
    80; United States in, 86–87, 144–45, 149;
    urban space along, 63; worlding in, 112–13,
    146–50
Pacific Rim (film), 41
Pak Chan-uk, 150
Palmer, Julius A., 151
Papua New Guinea, 78, 147, 149
Parasite (film), 150
Pascal, Blaise, 76
pastoral, 33, 118–19
pedagogy, 111–15
Pendell, Dale, 7, 11
Penitence (Shōda), 140
People's Republic of China. *See* China
Peralta, Stacy, 15
Perlman, Michael, 136
Perloff, Marjorie, 142
Personal Matter, A (Ōe), 134–35
Philip, M. NourbeSe, 72–73
Philippines, 63, 73–74, 82, 106, 146–47
pidgin artists, 80–81
Pidgin Eye (Balaz), 142–43
Pirates of the Caribbean (film), 171n59
Place in Space, A (Snyder), 85, 88, 120
place/place-making, 25–27, 38–39, 51–54,
    56–65, 73–76, 84–91, 94–102
planetary beatitude, 123–24

planetary being, 42
planetary capitalism, 53, 57
planetary endgame, 54
planetary global Oceania, 157
planetary nexus, 1–2
planetary nourishment, 4
planetary peril, 26
planetary turns, 13
planetary world, 57
Plastic Ocean (Moore, C.), 20
Poe, Edgar Allan, 46
Poeta en San Francisco (Reyes), 104
poetics: becoming oceanic, 115; biopoetics, 7;
    blue, 23; blue cultural, 14; blue ecopoet-
    ics, 24–25; ecopoetics, 32–33; of Oceania,
    39–41, 43–50, 86; poetic practices, 61;
    of resistance, 2; of reworlding, 157–59;
    of Spahr, 37–38; sublime and, 16–18;
    of totality, 54; transpacific ecopoetics,
    32–33; worlding, 124–25. *See also*
    ecopoetics
Poetry Deal, The (di Prima), 161–62
Poland, 93, 98
"Polynesian Hong Kong" (Balaz), 83
Pompallier, Jean Baptiste François, 76
postcolonialism, 64, 78–79, 83–84, 88,
    99–102, 187n25, 195n36
di Prima, Diane, 111, 120–21, 161–62, 166
Probyn, Elspeth, 4
Pule, John, 82
"Pulling of Olap's Canoe, The" (Reuney), 91
Purdy, Jebediah, 67

race/racism. *See specific topics*
Radhakrishnan, Radha, 61
Rankine, Claudia, 23
"Rant" (di Prima), 111, 121
Raporgan, Shyman, 83–84
Reagan, Ronald, 102–3
refugees, 92–94, 97–103, 114–15. *See also*
    migrants
Regarding Wave (Snyder), 85
Reid, Whitelaw, 35
Reimagining the American Pacific (Wilson),
    73
religion. *See* specific religions
"Requiem for the American Empire" (Vidal),
    126
Reuney, Theophil Saret, 91
"Revolutionary Letter #2" (di Prima),
    161–62
Revolutionary Letters (di Prima), 161–62

reworlding: deworlding and, 26, 57, 60–67,
    80–81, 100, 157; ecology, 160; global cities,
    86–87; Oceania, 38–39; philosophy of, 8;
    poesis of, 49–50; poetics of, 157–59; for
    survival, 56–57; unworlding and, 53
Rexroth, Kenneth, 25, 80, 116
Reyes, Barbara Jane, 104
Rice, Doyle, 67
Riding Giants (film), 15
Rip Rap (Snyder), 120
Roberts, Brian Russell, 3, 162
Robinson, Kim Stanley, 141, 169
Roy, Ananya, 62–63, 64
Rukeyser, Muriel, 86
Rushdie, Salman, 102
Russia, 6, 32, 80, 88, 102, 128, 134

Sacks, Sam, 46
Said, Edward, 61, 74
Saijo, Albert, 85, 90, 91
Sakaki, Nanao, 84–85, 90–91
Salinger, J. D., 23
Salt Wind, The / Ka Makani Pa'akai (McDou-
    gall), 14–15, 38
Samoa (Ehlers), 21
Sampas, Sebastian, 45
Sanders, Rupert, 6
San Francisco: Asia Pacific and, 80; Beat cul-
    ture in, 120, 122–23; City Lights Books in,
    1–2; coastal California and, 8; culture of,
    11–12, 111–12; New York City and, 60–61,
    108; Pacific Rim and, 25; politics in, 115–17;
    refugees in, 97; Santa Cruz and, 113–15;
    Silicon Valley and, 12–13, 116–17; Snyder
    on, 88–89, 120–22; University of San Fran-
    cisco, 86; urban space in, 25–26; worlding,
    92–93, 103–8, 117–19
Santa Cruz, 2, 12, 26–27, 32–34, 73,
    111–15
Santayana, George, 19
Santos Perez, Craig, 36, 39, 85–87, 89
Sassen, Saskia, 96–97
Schalansky, Judith, 22
Schmitt, Carl, 20–21
Schonfelder, Kristin, 10–11
Schultz, Susan, 164
"Sea" (Kerouac), 48–49
Sea around Us, The (Carson), 48
Sea Is My Brother, The (Kerouac), 45–48
"Searching for Nei Nim'anoa" (Teaiwa),
    158–59
sea slaves, 15–19, 96–97, 171n58

Seaspiracy (documentary), 53

Sekula, Allan, 16, 40

Selkirk, Alexander, 22

Shark That Ate the Sun, The (Pule), 82

Shell, Marc, 23

Shepard, Benjamin, 12

Shepard, Thomas, 74

Sherry, Michael, 136–37

Shōda, Shinoe, 140

"Shrinking the Pacific" (Inada), 82–83

Shurin, Aaron, 86

Silk Road, 5, 18, 31–32, 54–55

Singapore, 43, 52, 64–65, 147

Slat, Boyan, 24

Sloterdijk, Peter, 17–18, 22, 41–42

Smithsimon, Greg, 12

Snowden, Edward, 102

Snowpiercer (film), 55, 93, 150

Snyder, Gary, 47; on ecology, 76–77, 84–85;
    Kerouac and, 122–23; reputation of, 32, 36,
    88–91, 105, 119, 163, 166; on San Francisco,
    88–89, 120–22

Snyder, Masa, 84–85

Solnit, Rebecca, 61, 104, 119

someone's dead already (Eisen-Martin),
    25–26

Sorrows of Empire, The (Johnson), 149

South Korea: business in, 63; China and,
    73–74; Hawai'i and, 25; Japan and, 36;
    Korean War, 2; Korea University, 11–12;
    New Zealand and, 72; North Korea and, 55,
    93; Pacific Rim and, 125, 143–46; Taiwan
    and, 142, 165; trade with, 192n14; urban
    space in, 65

Spahr, Juliana, 7–8, 14–15, 36–38, 80–81, 114,
    163–64

"Spear Fisher" (Balaz), 89

Spicer, Jack, 23, 163

Spillers, Hortense, 71

Spivak, Gayatri, 87–88, 149

Splay Anthem (Mackey), 92–93

"Spumante" (Channer), 43–44

Starsdown (Bernes), 7

Star Waka (Sullivan), 14–15, 89

Stein, Murray, 16

Steinbeck, John, 20

Stevens, Wallace, 54, 61, 105

Stevenson, Robert Louis, 20

sublime: blue poetics as, 23; Hiroshima,
    126–30; imaginary, 41–42; Lapham on, 20;
    nuclear, 132–40, 190n4, 191n6; Pacific sub-
    limity, 78; poetics and, 16–18; Romanticism

and, 32; with technology, 191n8; Whitman
    on, 171n62

Sullivan, Robert, 14–15, 86, 89

Superman IV (film), 129

Sympathy for Mr. Vengeance (Park), 150

sympoesis, 5, 53–54, 67, 72

Syria, 96–97

Tabrizi, Ali, 53

Tagore, Rabindranath, 25

Taiwan: Australia and, 43; China and, 93–94,
    107; coastal California and, 124–25; ecology
    of, 25; geography of, 3–4; in globalization,
    33; Hawai'i and, 98; Huang on, 90; Japan
    and, 72, 84; Philippines and, 73–74; postco-
    lonialism in, 83–84; South Korea and, 142,
    165; transpacific, 119

Talbot, Joe, 108

Tales of the Tikongs (Hau'ofa), 81

Taylor, Bayard, 116

Tea, Michelle, 104

Teaiwa, Teresia, 14, 82, 156–60, 162, 166,
    184n46, 194n29

Tempest, 17

Te Ponga, Alice, 4

Thank God for the Atom Bomb (Fussell),
    136

Thorne, Christian, 21

Through Other Continents (Wai Chee Di-
    mock), 39

Thurston, Lorrin A., 151

Tinfish (Santos Perez), 39

Titanic (ship), 17

Tjibaou, Jean-Marie, 89

Tonouchi, Lee, 142, 165

Torda, Greg, 67

Torro, Guillermo del, 41

transfiguration, 11, 13–15, 78–80

Trend, David, 59, 121

Tripmaster Monkey (Kingston), 104–5, 107–8,
    123–24

Trouble No More, 163

Trout Fishing in America (Brautigan), 114–15,
    119–20

Truman, Harry, 132

Trump, Donald, 6, 55, 93–98, 188n51

Tsing, Anna, 5, 27, 56–57

Turkey, 96–97

Turtle Island (Snyder), 85

Twain, Mark, 77–78

Twenty Thousand Leagues under the Sea
    (Verne), 15, 170n53

United Nations, 7, 97, 175n37

United States: activism in, 107; American Orientalism, 148–49, 193n15; Anthropocene for, 55; as archipelagic nation, 3–4; Asians in, 103–4; Atlantic Ocean and, 16–20, 32–34, 43–45; California to, 46; China and, 22, 33–34, 95, 104–5, 173n4, 174n30, 176n43; after Cold War, 147–48; Connecticut and, 163–66; diaspora of, 27; Europe and, 87, 93, 95–96; in globalization, 97–98, 126–32; Guam and, 39, 43; in Gulf War, 127–28; Hawai'i and, 37, 150–56; hegemony of, 5–6, 60–61, 150–51, 173n8; ideology, 132; Japan and, 7, 34, 41, 123, 136–40, 192n14; Manifest Destiny in, 162; Mexico and, 47, 94, 98, 102–3; as nuclear empire, 58; ocean commons to, 34–35; in Oceania, 34–36; Pacific Ocean to, 35–36, 43; in Pacific Rim, 86–87, 144–45, 149; Philippines and, 146–47; in postcolonialism, 144; Russia and, 32, 102, 134; sanctuary cities in, 103; Sweden and, 96; in war, 2, 102; waste in, 24; in World War II, 45

unworlding, 53, 56

Unwritten Literature of Hawai'i (Emerson, N.), 155

Urbain, Jean-Didier, 13

Vaneigem, Raoul, 49–50

Verne, Jules, 15, 170n53, 171n59

Vidal, Gore, 126

Vietnam War, 2, 105

"Visons of Johana," 117

Vivekanand, Babu, 81

Waihe'e, Jennifer, 165

Walker, Richard, 73, 112, 118–19

Wall Street Journal, 47

Wanderlust (Solnit), 61

Wang, Chih-ming, 107

Wang, June, 56

Warhol, Andy, 145–46

Wark, McKenzie, 8–9, 55

Warner, Marina, 171n59

Waters, Lindsay, 144

We Are the Ocean (Hau'ofa), 77–78, 90

Welch, Lew, 105, 123

Well Then There Now (Spahr), 14–15

Wendt, Albert, 40

Westlake, Wayne, 80–81

Whalen, Philip, 123

"What the Sea Throws Up at Vlissingen" (Ginsberg), 23–24

Whetu Moana (anthology), 89

Whitman, Walt, 16–17, 46, 137, 155, 171n62

Williams, Raymond, 118–19

Winchester, Simon, 20

Winduo, Steven Edmund, 78

Wolfe, Alan, 190n4

Wolfe, Thomas, 46

Woo-Cumings, Meredith, 149

Woolf, Virginia, 40

Words in Motion (Gluck and Tsing), 57

"World Bank English," 81

worlding: Asia Pacific, 51–60, 71–80, 177n2; as critical practice, 113–14; definitions of, 27; deworlding and, 55–56; with ecumene frame, 76; ethics of, 60; ethos, 61; in globalization, 148; by Heidegger, 179n28; migration and, 24–25; narratives, 108; in Pacific Rim, 112–13, 146–50; poesis, 4–5; poetics, 124–25; poesis, 26; San Francisco, 92–93, 103–8, 117–19; in Silicon Valley, 121–22; at University of California, Santa Cruz, 33; as world dwelling, 49–50; worlding poesis, 4–5

Worlding (Trend), 59

Worlding Cities (collection), 53, 62–65

Worlding Multiculturalisms (Goh), 64

Worlding Project, The (collection), 52–53, 57, 59–61, 64–65, 113

World War II. *See specific topics*

Wozniak, Steve, 117

Wright, Richard, 4

Xi Jinping, 18

Yamashita, Karen Tei, 82, 104–7, 123–24, 188n51, 188n57

Yeo, George, 176n43

You Must Change Your Life (Sloterdijk), 22

Yu-Fang Cho, 58

Zen space, 118–20

Žižek, Slavoj, 94

Zong! (Philip), 72–73

www.ingramcontent.com/pod-product-compliance
Lightning Source LLC
Chambersburg PA
CBHW020858270326
41928CB00006B/759